Schriftenreihe zur Physiologie und Biochemie der Pflanzen

Herausgeber: Prof. Dr. A. Schaller

Band 15/2023 Eric Robin Bühler

Regulation of stress responses by sulfated peptides in Arabidopsis

D 100 (Diss. Universität Hohenheim)

Regulation of stress responses by sulfated peptides in Arabidopsis

Dissertation to obtain the doctoral degree of Natural Sciences (Dr. rer. nat.)

Faculty of Natural Sciences

University of Hohenheim

Institute of Biology

submitted by

Eric Robin Bühler

from Lahr

2023

Bibliografische Information der Deutschen Nationalbibliothek
Die Deutsche Nationalbibliothek verzeichnet diese Publikation in der Deutschen Nationalbibliografie; detaillierte bibliographische Daten sind im Internet über http://dnb.d-nb.de abrufbar.
1. Aufl. - Göttingen: Cuvillier, 2024
Zugl.: Hohenheim, Univ., Diss., 2024

Dekan:	Prof. Dr. Jan Frank
1. berichtende Person:	Prof. Dr. Andreas Schaller
2. berichtende Person:	Prof. Dr. Andreas Hiltbrunner
3. prüfende Person:	Prof. Dr. Waltraud Schulze
Eingereicht am:	15.09.2023
Mündliche Prüfung am:	19.12.2023

Die vorliegende Arbeit wurde am 04.12.2023 von der Fakultät Naturwissenschaften der Universität Hohenheim als „Dissertation zur Erlangung des Doktorgrades der Naturwissenschaften“ angenommen.

Nonnenstieg 8, 37075 Göttingen
Telefon: 0551-54724-0
Telefax: 0551-54724-21
www.cuvillier.de

1. Auflage, 2024
Gedruckt auf umweltfreundlichem, säurefreiem Papier aus nachhaltiger Forstwirtschaft.
ISBN 978-3-7369-7948-2
eISBN 978-3-7369-6948-3

Content

Abbreviations

AAO3 - ABSCISIC ALDEHYDE OXIDASE

ABA - abscisic acid

ALA - 5-aminolevulinic acid

ALE1 - ABNORMAL LEAF SHAPE1

APL3 - ADP-Glucose Pyrophosphorylase large subunit

BAK1 - BRI1-associated receptor kinase 1

BAM3 - BARELY ANY MERISTEM 3

bHLH - basic helix-loop-helix

BR1 - BRASSINOSTEROID INSENSITIVE 1

BZR1 - BRASSINAZOLE RESISTANT1

CAB - CHLOROPHYLL A/B BINDING PROTEIN

CaMV - cauliflower mosaic virus

CAPE - CAP-DERIVED PEPTIDE

CBB - coomassie brillant blue

CEP - C-terminally encoded peptide

CHI - CHALCONE ISOMERASE

CHS - CHALCONE SYNTHASE

CIF- CASPARIAN STRIP INTEGRITY FACTOR

CLE - CLAVATA3(CLV)/EMBRYO-SURROUNDING REGION RELATED

CLEL - CLE-like

CLV - CLAVATA

COP1 - CONSTITUTIVE PHOTOMORPHOGENESIS 1

DAMP - damage-associated molecular pattern

DFR - DIHYDROFLAVONOL 4-REDUCTASE

ELF3 - EARLY FLOWERING3

ELIP - EARLY LIGHT-INDUCABLE PROTEIN

EPI - extracellular proteinase inhibitor

ESI-MS - electrospray ionization mass spectrometry

FC1 - FERROCHELATASE1

FLS2 - FLAGELLIN SENSITIVE2

FLU - FLUORESCENT IN BLUE LIGHT

GC-MS - gas chromatography–mass spectrometry

GFP - green fluorescent protein

GLK - GOLDEN-2-LIKE

GLV - GOLVEN

GPT2 - glucose 6-phosphate/phosphate translocator2

GSO - GASSHO

GUN - GENOMES UNCOUPLED

HFR1 - LONG HYPOCOTYL IN FAR-RED

HPAT - hydroxyproline o-arabinosyltransferase

HPLC - high performance liquid chromatography

HRP - horseradish peroxidase

HY5 - ELONGATED HYPOCOTYL 5

IAA - indole-3-acetic acid

IDA - INFLORESCENCE DEFICIENT IN ABSCISSION

IDL - IDA-like

JAZ - JASMONATE ZIM-DOMAIN

LC-MS - liquid chromatography mass spectrometry

LDOX - LEUCOANTHOCYANIDIN OXYGENASE

LHC - light-harvesting chlorophyll a/b-binding

Linc - lincomycin

LRR-RLK - leucin-rich repeat receptor-like kinase

MAMP - microbe-associated molecular pattern

MKK - mitogen-activated map kinase kinase

MPK - MITOGEN-ACTIVATED PROTEIN KINASE

MS - mass spectrometry

NTC - non-template control

ORF - open reading frame

P4H - proline-4-hydroxylase

PA - protease-associated

PAMP - PATHOGEN-ASSOCIATED MOLECULAR PATTERN

PAP - PRODUCTION OF ANTHOCYANIN PIGMENT

PAPs - 3'-phosphoadenosine 5'-phosphosulfate

Pchlide - protochlorophyllide

PEP - PLANT ELICITOR PEPTIDE3

phyB - phytochrome B

Phyt2 - phytaspase-2

PIF - PHYTOCHROME INTERACTION FACTOR

PIP - PAMP-INDUCES PEPTIDES

PLT - PLETHORA

POR - protochlorophyllide oxidoreductase

PR-1b - PATHOGENESIS-RELATED PROTEIN1b

PSK - PHYTOSULFOKINE

PSKR - phytosulfokine receptor

PSY - PLANT PEPTIDE CONTAINING SULFATED TYROSINE

PTI - pattern triggered immunity

PTM - post-translational modification

RALF - RAPID ALKALANIZATION FACTORS

RFU - relative fluorescence units

RGF - ROOT GROWTH FACTOR

RGI - ROOT MERISTEM GROWTH FACTOR 1 INSENSITIVE

RITF1 - RGF1-INDUCIBLE TRANSCRIPTION FACTOR

ROS - reactive oxygen species

SBT - subtilisin-like serine protease (subtilase)

SCOOP - SERINE-RICH ENDOGENOUS PEPTIDES

SDS - sodium dodecylsulfate

SDS-PAGE - sodium dodecyl sulfate–polyacrylamide gel electrophoresis

SERK - SOMATIC EMBRYOGENESIS RECEPTOR-LIKE KINASE

SPA - SUPPRESSOR OF PHYA-105

SPMP - small posttranslational modified peptides

SUC1 - sucrose transporter1

TEM - transmission electron microscopy

TLR - Toll-like receptors

TPST - tyrosylprotein sulfotransferase

TUB4 - TUBULIN4

TWS1 - TWISTED SEED1

UBP - ubiquitin-specific proteases

UF3GT - UDP-GLUCOSE:FLAVONOID 3- O-GLUCOSYLTRANSFERASE

YDA - YODA

Zusammenfassung

Kleine posttranslational modifizierte Peptide nehmen eine bedeutende Rolle in der Pflanzenentwicklung und dem Wachstum ein. Sie werden als inaktive Vorläuferproteine synthetisiert und durch mehrere posttranslationale Modifikationen gereift. Zu diesen Modifikationen zählen die Prolin-Hydroxylierung, Glykosylierung, Tyrosin-Sulfatierung und proteolytische Prozessierungen. Neueste Erkenntnisse haben auch wichtige Rollen für Peptide in der pflanzlichen Stressantwort aufgedeckt. Pflanzen sind aufgrund ihrer sessilen Lebensweise abiotischem und biotischem Stress ausgesetzt und müssen die Waage zwischen Stressantwort und Wachstum balancieren um zu überleben. Dieser Prozess wird folglich von vielen Faktoren reguliert. Vor Allem die Untergruppe von tyrosin-sulfatierten Peptiden bekommt hierbei eine immer größere Bedeutung. Diese Modifikation ist in den meisten Fällen essentiell für die Funktion der Peptide, da Sie wichtig für die Rezeptorbindung oder für weitere posttranslationale Modifikationsschritte ist. Im Rahmen dieser Arbeit wurden die Funktionen von sulfatierten Peptiden in Arabidopsis erforscht. Das sulfatierte Peptid CLE-LIKE6 (CLEL6) wurde als negativer Regulator der Anthocyan-Biosynthese in Arabidopsis identifiziert. Die Biogenese von CLEL6 hängt von der proteolytischen Prozessierung durch die Subtilisin-ähnliche Serinprotease6.1 (SBT6.1) und von der Tyrosinsulfatierung durch das Enzym tyrosylprotein sulfotransferase (TPST) ab. *Loss-of-function* Mutanten von *sbt6.1* oder *tpst*, welche kein aktives CLEL6 produzieren können, zeigten eine erhöhte Akkumulierung von Anthocyanen unter Lichtstress. Diese Überakkumulation von Anthocyanen in den Mutanten konnte durch die Applikation von reifem CLEL6 komplementiert werden. Zusätzlich inhibierte eine Überexpression und externe Applikation von CLEL6 die Expression von Genen, die an der Anthocyan-Biosynthese beteiligt sind. Des Weiteren konnte durch einen genetischen Ansatz, unter Verwendung einer fünffach-Mutante der ROOT MERISTEM GROWTH FACTOR 1 INSENSITIVE (RGI) Rezeptorfamilie, eine essentielle Funktion der RGI-Rezeptorfamilie im CLEL6-Signalweg gezeigt werden. Darüber hinaus konnte gezeigt werden, dass die CLEL6 Expression durch Licht reduziert wird. Die Daten legen nahe, dass CLEL6 die Anthocyan-Biosynthese in etiolierten Keimlingen inhibiert und durch Lichteinwirkung die *CLEL6* Expression reduziert wird, wodurch die inhibitorische Funktion von CLEL6 entkräftet und die Anthocyan-Biosynthese aktiviert wird.

Des Weiteren wurde das sulfatierte Peptid Phytosulfokin (PSK) als Regulator von Trockenstress identifiziert. Die Expression von Genen, die für die Vorläuferpeptide von PSK kodieren, sowie Transkripte von drei subtilisin-ähnlichen Serinproteasen, SBT1.4, SBT3.7 und SBT3.8, wurden als Reaktion auf osmotischen Stress hochreguliert. Anschließend wurde gezeigt, dass die SBT3.8 an der Biogenese des bioaktiven PSK-Peptids beteiligt ist. Der PSK1-Vorläufer (ProPSK1) wurde von SBT3.8 am C-Terminus des PSK-Pentapeptids Aspartat-spezifisch gespalten. Erzeugte *loss-of-function* Mutanten von *sbt3.8* wiesen eine niedrigere Resistenz gegenüber osmotischem Stress auf, welche durch Zugabe von reifem PSK komplementiert werden konnte. Des Weiteren wurde die osmotische Stressresistenz bei Pflanzen, die proPSK1 oder SBT3.8 überexprimierten, verbessert, was zu einem höheren Frischgewicht und einer verbesserten lateralen Wurzelentwicklung bei transgenen Pflanzen im Vergleich zu Wildtyp-Pflanzen führte. Die Daten legen nahe, dass die Expression von *PSK* und *SBT3.8* bei osmotischem Stress erhöht wird, um die Stresstoleranz zu steigern.

Sulfatierte Peptide haben auch eine wichtige Rolle in der Entwicklung der Pflanze. Etiolierte Keimlinge der *tpst* Mutante, welche als *loss-of-function* Mutante für alle sulfatierten Peptide gesehen werden kann, zeigten einen verfrühten Photomorphogenese-Phänotyp. Dieser äußerte sich durch ein verkürztes Hypokotyl, geöffnete Kotyledonen, eine hohe Akkumulation von Protochlorophyllid, erhöhte Transkriptmengen mehrerer mit der Photomorphogenese zusammenhängender Gene und gestörte Etioplastenbildung. Der verfrühte Photomorphogenese Phänotyp der *tpst*-Mutante wird vermutlich durch eine erhöhte Expression des Photomorphogenese-induzierenden Transkriptionsfaktors GOLDEN-2-LIKE (GLK1) verursacht. Ein weiteres Indiz hierfür ist ein ähnliches Verhalten bei Überexpression von GLK1 (*35S:GLK1*) und der *tpst*-Mutante. Beide können unter aktivierter Hemmung der Photomorphogenese durch *retrograde signaling*, die durch die Behandlung mit Lincomycin ausgelöst wurde, de-etiolieren. Des Weiteren konnte durch Transmissionselektronenmikroskopie von *tpst* Etioplasten, eine erhöhte Akkumulierung von Stärke in den Plastiden festgestellt werden, was eine Rolle des TPST Enzyms in der Regulierung der Stärkeakkumulation vermuten lässt.

Summary

Small posttranslationally modified peptides play a significant role in plant development and growth. They are synthesized as inactive precursor proteins and matured through multiple posttranslational modifications. These modifications include proline hydroxylation, glycosylation, tyrosine sulfation, and proteolytic processing. Recent findings have also uncovered important roles for peptides in plant stress responses. Plants, due to their sessile lifestyle, are exposed to abiotic and biotic stress and must balance stress response and growth to survive. Consequently, this process is regulated by many factors. Particularly, the subgroup of tyrosine-sulfated peptides is gaining increasing importance. This modification is essential for peptide function in most cases, as it is crucial for receptor binding or further posttranslational modification steps. In this study, the functions of sulfated peptides in Arabidopsis were investigated.

The sulfated peptide CLE-LIKE6 (CLEL6; e.g. RGF6/GLV1) was identified as a negative regulator of anthocyanin biosynthesis in Arabidopsis. The biogenesis of CLEL6 depends on the proteolytic processing by the subtilisin-like serine protease 6.1 (SBT6.1) and tyrosine sulfation by the enzyme tyrosylprotein sulfotransferase (TPST). Loss-of-function mutants of *sbt6.1* or *tpst*, which cannot produce active CLEL6, showed increased accumulation of anthocyanins under light stress. This overaccumulation of anthocyanins in the mutants could be complemented by the application of mature CLEL6. Additionally, overexpression and external application of CLEL6 inhibited the expression of genes involved in anthocyanin biosynthesis. Furthermore, a genetic approach using a quintuple mutant of the ROOT MERISTEM GROWTH FACTOR 1 INSENSITIVE (RGI) receptor family demonstrated an essential function of the RGI receptor family in the CLEL6 signaling pathway. Moreover, it was shown that CLEL6 expression is reduced by light. The data suggest that CLEL6 inhibits anthocyanin biosynthesis in etiolated seedlings and that CLEL6 expression is reduced by light exposure, thus counteracting the inhibitory function of CLEL6 and activating anthocyanin biosynthesis.

Furthermore, the sulfated peptide phytosulfokine (PSK) was identified as a regulator of drought stress. The expression of genes encoding the precursor peptides of PSK, as well as transcripts of three subtilisin-like serine proteases, SBT1.4, SBT3.7, and SBT3.8, were upregulated in response to osmotic stress. Subsequently, it was shown that SBT3.8 is involved in the biogenesis of the bioactive PSK peptide. The PSK1 precursor (ProPSK1) was specifically cleaved at the C-terminus of the PSK pentapeptide by SBT3.8 in an aspartate-specific manner. Loss-of-function mutants of *sbt3.8* exhibited lower resistance to osmotic stress, which could be complemented by the addition of mature PSK. Furthermore, osmotic stress resistance was improved in plants overexpressing proPSK1 or SBT3.8, leading to higher fresh weight and improved lateral root development compared to wild-type plants. The data suggest that the expression of PSK and SBT3.8 is increased during osmotic stress to enhance stress tolerance.

Sulfated peptides also play an important role in plant development. Etiolated seedlings of the *tpst* mutant, which can be considered a loss-of-function mutant for all sulfated peptides, exhibited an early photomorphogenesis phenotype. This was characterized by a shortened hypocotyl, open cotyledons, high accumulation of protochlorophyllide, increased transcript levels of several genes associated with photomorphogenesis, and disrupted etioplast formation. The early photomorphogenesis phenotype of the *tpst* mutant is presumed to be caused by increased expression of the photomorphogenesis-inducing transcription factor GOLDEN-2-LIKE (GLK1). Further evidence for this is the similar behavior of GLK1 overexpressors (35S:GLK1) and the *tpst* mutant. Both can de-etiolate under activated inhibition of photomorphogenesis triggered by lincomycin treatment. Furthermore, transmission electron microscopy of *tpst* etioplasts revealed increased starch accumulation in plastids, suggesting a role of the TPST enzyme in regulating starch accumulation.

1. Introduction

1.1 Balance between plant growth and stress responses

Due to the phenomenon of climate change and thus altered environmental conditions for plants, the understanding of the plant's stress responses is moving into the focus of vast scientific studies. These responses are triggered by abiotic stresses such as salinity, drought, and heat (Zhang et al., 2022), or biotic stresses, for instance, caused by insects, fungi, or bacteria (Verma et al., 2016). The plant stress response involves a delicate balance between growth and stress defense/tolerance, ultimately influencing the plant's fitness and survival (Figure 1).

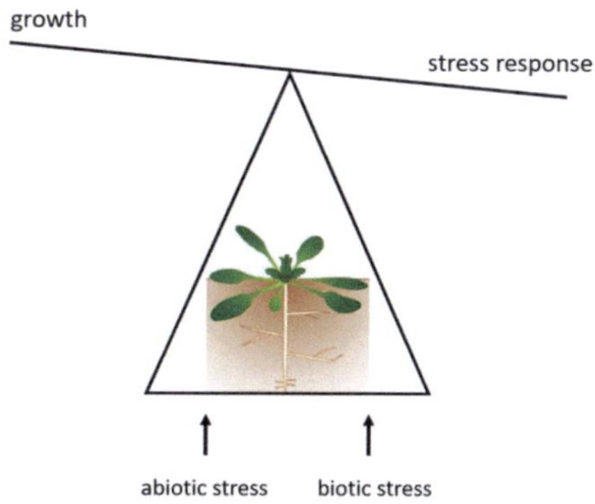

Figure 1. Balancing plant growth and stress signaling. Plants need to adapt upon abiotic and biotic stress. Thus, the plant needs to keep the balance between growth and stress response.

Morphologic changes observed in stressed plants include reduction of growth, senescence, early flowering, stomatal closure, and many mechanisms that enhance the stress tolerance (Zhang et al., 2022). Due to the importance for plant survival, plant stress response has to be regulated by many factors. Well-established regulators of plant stress response are the phytohormones, including abscisic acid (ABA), ethylene, cytokinin, auxin, gibberellin, jasmonate, salicylic acid, and strigolactones (Waadt et al., 2022). For instance, during osmotic stress triggered by salinity or drought, ABA adjusts growth and induces the closure of the stomata to reduce water deficit (Bharath et al., 2021; Verma et al., 2016). Auxin regulates hydropatterning, a mechanism that induces lateral root formation close to water, consequently increasing the water supply during low water availability (Orosa-

Puente et al., 2018). Root hydrotropism, the root growth towards moist soil, is regulated by asymmetric distribution of cytokinin in the root tip (Chang et al., 2019). Furthermore, auxin and brassinosteroids mediate the hyponastic response to elevated temperatures (Delker et al., 2021), and gibberellins regulate plant growth under cold stress (Lantzouni et al., 2020). During biotic stress, salicylic acid, ethylene, and jasmonate primarily mediate plant immunity (Bari et al., 2008). In addition to phytohormones, small posttranslational modified peptides, also known as peptide hormones, have emerged as regulators of plant stress responses (Kim et al., 2021, Hou et al., 2021).

1.2 Maturation of peptide hormones includes several steps of modifications

Plant peptides can be derived either from larger precursor proteins or transcribed directly from small open reading frames (Tavormina et al., 2015). Small posttranslational modified peptides (SPMPs) are classified into two main categories: cysteine-rich peptides and post-translationally modified peptides (Matsubayashi, 2014). Cysteine-rich peptides are characterized by a high number of cysteine residues, which enables the formation of disulfide bridges and thereby impacts on the folding and stability of the peptides. For instance, RAPID ALKALANIZATION FACTORS (RALFs) that regulate pollen tube growth and salt tolerance (Somoza et al., 2020; Zhao et al., 2018). LURE peptides, on the other hand, are identified as pollen tube attractants, thereby playing a crucial role in plant reproduction (Zhong et al., 2019).

Posttranslational modified peptides are synthesized as larger pre-pro-peptides that contain a N-terminal signal peptide. This signal peptide recruits the precursor into the secretory pathway and is cleaved upon entry. The biogenesis of posttranslationally modified peptides can involve several steps including proteolytic cleavage from the precursor, tyrosine sulfation, proline hydroxylation, and/or arabinosylation of hydroxyprolines (Figure 2). Many of these modifications are essential for the functional activity of the mature peptides (Stührwohldt and Schaller 2019). Once matured, the peptide is secreted into the apoplast and recognized by specific receptors, which initiate downstream signaling. Leucine-rich repeat receptor-like kinases (LRR-RLKs) are commonly involved as receptors in posttranslational modified peptide perception. Each stage of peptide maturation and perception will be described further in the subsequent sections.

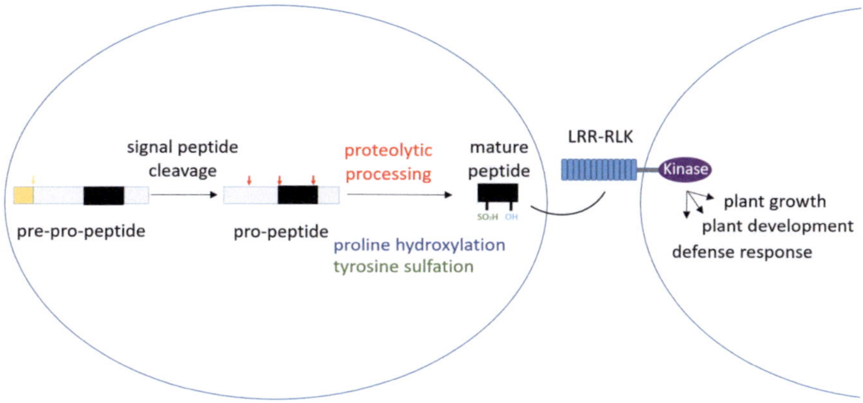

Figure 2. Simplified scheme of the posttranslational modification of peptides. The larger pre-pro-peptide is cleaved to release the signal peptide (yellow) upon entering the secretory pathway. Proteolytic processing events (red) done by subtilases cleave the peptide to its mature length. Tyrosine sulfation (green) is catalysed by the TPST enzyme, while proline hydroxylation (blue) is catalysed by P4Hs. The mature peptide is released into the apoplast and is perceived by a receptor (LRR-RLK) which can trigger effects on plant growth, plant development or defence responses.

1.3 Modification and perception of peptide hormones

1.3.1 Proteolytic processing by subtilases

The proteolytic cleavage of posttranslationally modified peptide precursors is mediated by enzymes called subtilases, which are encoded by 56 genes in Arabidopsis (Rautengarten et al., 2005). Due to redundancy, the specific functions of many subtilases have yet to be discovered. The term "subtilases" originates from subtilisin, a protease found in *Bacillus subtilis*, and their catalytic activity is defined by the arrangement of Aspartate, Histidine, and Serine residues, known as the catalytic triad (Smith et al., 1966; Meyer et al., 2016). Plant subtilases are synthesized as pre-proenzymes and undergo several activation steps, similar to their ligands, the SPMPs (Meyer et al., 2016). The structure of subtilases comprises a signal peptide, prodomain, subtilase domain, PA domain, and FnIII-like domain (Schaller et al., 2018). The signal peptide guides the subtilase into the secretory pathway and is cleaved upon entry (Schaller et al., 2018). The prodomain is necessary for proper folding and inhibits subtilase activity until it reaches its destination. The prodomain is subsequently cleaved autocatalytically in a pH-dependent manner upon entering the acidic trans-Golgi (Meyer et al., 2016). The subtilase domain contains the catalytic triad and is responsible for cleaving the subtilase's substrate. The PA domain serves as a protein-protein interaction domain (Mahon et al., 2000). Subtilase processing can either involve pre-processing of the precursor to facilitate correct folding or cutting the peptide into its final length. For example, the SPMP CLAVATA3/ESR-RELATED (CLE40), which regulates stem cell differentiation, is processed by SBT1.4, SBT1.7, and SBT4.13 (Stührwohldt et al., 2020a). The INFLORESCENCE DEFICIENT IN ABSCISSION (IDA) peptide, involved in floral organ abscission, is processed by SBT4.12, SBT4.13, and SBT5.2 (Schardon et al., 2016). In tomato, the wound response-regulating peptide systemin is cleaved by phytaspase1 and phytaspase2 (Beloshistov et al., 2018). To address the redundancy of subtilases, the expression of subtilase-specific inhibitor proteins from *Phytophthora infestans*, such as EPI1 (extracellular proteinase inhibitor 1) and EPI10, under the control of tissue-specific promoters, has become an important tool for analysing peptide functions. This approach has facilitated the identification of subtilases involved in IDA maturation (Schardon et al., 2016).

1.3.2 Proline hydroxylation by P4Hs

Proline hydroxylation is a common PTM and catalyzed by proline-4-hydroxylases (P4H), which are encoded by 13 genes in Arabidopsis (Hieta et al., 2001; Velasquez et al., 2015). Besides proline, the catalyzed reaction includes oxygen (O_2) and 2-oxoglutarate as co-substrates, Fe^{2+} as a co-factor and ascorbate for full activity (Kivirikko et al., 1989; Annunen et al., 1998). The Fe^{2+} ions are bound by a conserved histidine-x-aspartate motif in the P4Hs (Yamaguchi et al., 1998; Yuasa et al., 2004). Proline hydroxylation plays a crucial role in the function of peptides. For instance, CLE40 proline hydroxylation prevents further cleavage events in the mature peptide sequence during the proteolytic processing of the CLE40 precursor, thus ensuring correct maturation (Stührwohldt et al., 2020a). Proline hydroxylation of IDA was found to be essential for receptor binding (Santiago et al., 2016). In contrast to the proteolytic processing by subtilases, proline hydroxylation is not always essential for the function of the peptides. The ROOT GROWTH FACTOR6 (RGF6) or RGF4 peptides are also active in the absence of their proline hydroxylation (Whitford et al., 2012; Ghorbani et al., 2016). Furthermore, proline hydroxylation serves as a basis for modifications like O-linked glycosylation, which is crucial for receptor binding (Amano et al., 2007). Golgi-located HPAT, encoded by three genes in Arabidopsis, catalyze the O-linked glycosylation (Ogawa-Ohnishi et al., 2013). O-linked glycosylation was identified as a critical factor for the functional activity of the sulfated peptide PLANT PEPTIDES CONTAINING SULFATED TYROSINE (PSY) (Amano et al., 2007).

1.3.3 Tyrosine sulfation by TPST

Tyrosine sulfation is a crucial modification that is catalyzed by the enzyme tyrosylsulfoprotein transferase (TPST). A gene encoding for the TPST enzyme is observed in both plant, and animal genomes (Moore et al., 2009; Komori et al., 2009). TPST utilizes the co-substrate 3'-phosphoadenosine 5'-phosphosulfate (PAPS) to transfer a sulfate group to the hydroxyl group of tyrosine (Moore et al., 2009). The animal genome encodes for two *TPST* genes. Both are classified as type II transmembrane proteins and are located in the trans-Golgi. Tyrosine sulfation is widespread in animals and has been associated with important roles in cellular processes. Notably, a *tpst1/tpst2* double mutant in mice leads to severely impaired postnatal survival (Huttner et al., 1982; Westmuckett et al., 2008). In contrast, the Arabidopsis genome encodes a single-copy gene for the TPST enzyme, which is classified as a type I transmembrane protein (Komori et al., 2009). *AtTPST* is widely expressed throughout the plant (Komori et al., 2009; Matsubayashi 2011). In plants, tyrosine sulfation mainly occurs in SPMPs and has been shown to be essential for the function of many peptides, particularly in receptor binding and proteolytic processing by subtilases (Amano et al., 2007; Meng et al., 2012; Ou et al., 2016; Doll et al., 2020; Royek et al., 2022; Stintzi and Schaller, 2022).

Thus far, four groups of sulfated peptides have been identified in Arabidopsis: phytosulfokine (PSK), PSYs, RGFs and CASPARIAN STRIP INTEGRITY FACTORs (CIFs) (Matsubayashi et al., 2014; Tavormina et al., 2014; Sauter et al., 2019). A knock out of the *AtTPST* gene results in a loss of function mutant for all sulfated peptides in Arabidopsis. The *tpst-1* mutant serves as a valuable tool for studying the function of these peptides. Initially, the *tpst-1* phenotype was described as dwarf plants accompanied with stunted roots, pale green leaves, reduced higher order veins, early senescence, and a reduced number of flowers and siliques (Komori et al., 2009). Further studies have revealed additional phenotypic characteristics of *tpst-1*, including hypersensitivity to fructose (Zhong et al., 2020), defects in embryonic cuticle formation (Doll et al., 2020; Giorgi et al., 2021), hypersensitivity to phosphate deficiency (Kang et al., 2014), defective maintenance of the root stem cell niche, and decreased meristematic activity (Zhou et al., 2010). Additionally, sulfotyrosine has been identified as an extracellular pH sensor (Liu et al., 2022). Thus far all sulfated peptides are perceived by receptors from the LRR-RLK family.

1.3.4 Peptide perception by LRR-RLKs

Leucine-rich repeat receptor-like kinases (LRR-RLKs) are receptors located in the plasma membrane that regulate various aspects of plant growth, development, and responses to abiotic and biotic stresses (Xu et al., 2016; Xi et al., 2019; Soltabayeva et al., 2022). In Arabidopsis, the LRR-RLK family consists of approximately 230 members (Xu et al., 2016). Structurally, they comprise an extracellular domain, a transmembrane domain, and a cytoplasmic kinase domain. Regarding the length of the extracellular domain, LRR-RLKs are divided into main or co-receptors (Xi et al., 2019). The large helical ectodomain of the main LRR-RLK binds its ligand by forming a heterodimer with a smaller ectodomain containing the co-receptor (Figure 3). In contrast, the human LRR-RLKs, known as Toll-like receptors (TLRs), homodimerize (Hothorn et al., 2016; Sameer et al., 2022). In plants, the co-receptor stabilizes the complex of the main LRR-RLK with the ligand and is required for kinase activation (Zhang et al., 2016). This triple complex is ligand-induced and ligand-dependent. Thus, in the absence of the ligand, the main- and co-receptor do not interact (Hohmann et al., 2017). Upon formation of the triple complex (ligand/receptor/co-receptor), the kinase domain of the main receptor, which determines downstream signaling, is activated by autophosphorylation and activates a cytoplasmic signaling cascade (Hohmann et al., 2017). The ligands of the LRR-RLK can include hormones, polypeptides or SPMPs (Zhang et al., 2016).

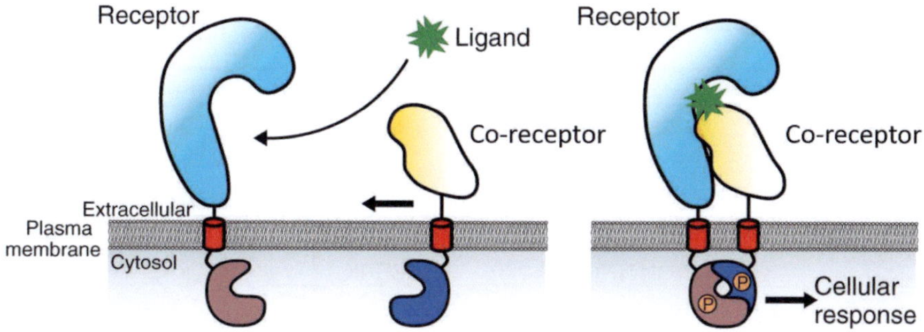

Figure 3. Main LRR-RLK forms a heterodimer with the ligand and co-receptor. Adapted from Brandt et al., 2016.

One of the most extensively studied LRR-RLKs is BRASSINOSTEROID INSENSITIVE 1 (BR1), which binds brassinosteroids using SOMATIC EMBRYOGENESIS RECEPTOR KINASE 3 (SERK3) as a co-receptor, thereby regulating numerous growth-related processes in plants (He et al., 2000; Santiago et al., 2013). SERK3 (also known as BAK1) also functions as a co-receptor for the FLAGELLIN SENSITIVE 2 (FLS2) receptor, which recognizes flagellin and triggers plant immunity (Chinchilla et al., 2007), highlighting the fact that the main LRR-RLK determines downstream signaling specificity. The SERK family comprises five members and has been shown to be essential for many processes related to plant growth, development, and immunity (Brandt et al., 2016). Furthermore, LRR-RLKs have been identified as receptors for SPMPs, as demonstrated e.g. by CLAVATA1 (CLV1), which binds the CLV3 peptide involved in shoot and floral meristem development, and HAESA, which binds the IDA peptide (Rojo et al., 2002; Cho et al., 2008). Recently, LRR-RLKs, including SERKs, have also been established as receptors for many sulfated peptides (Table 1).

Table 1. Perception of sulfated peptides by LRR-RLK.

Sulfated peptide	Receptor	Co-receptor	Publication
PSK	PSKR1, PSKR2	SERK3	Wang et al., 2015; Ladwig et al., 2015; Hartmann et al., 2014; Kaufmann et al., 2021
RGF1	RGI1, (RGI2, RGI3)	SERK	Ou et al., 2022
RGF5, RGF8	RGI	-	Fernandez et al., 2020
RGF7	RGI4, RGI5	SERK4, SERK5	Wang et al., 2021
RGF9	RGI3	SERK3	Stegmann et al., 2022
PSY1	PSY1R	-	Amano et al., 2007
PSY1-9	PSYR1, PSYR2, PSYR3	-	Ogawa-Ohnishi et al., 2022
CIF1, CIF2	GSO1, GSO2	-	Nakayama et al., 2017; Doublas et al., 2017
CIF3, CIF4	GSO1, GSO2		Truskina et al., 2022
TWS1	GSO1, GSO2	SERK1-3	Zhang et al., 2021

1.4 Sulfated peptides function as regulators of growth and development

1.4.1 PSK as growth and development regulator in various species

Sulfated peptides have gained increasing importance in plant stress responses. Among them, PSK has been extensively studied. Originally, PSK was discovered as a cell division-promoting factor in *Asparagus officinalis* cells (Matsubayashi et al., 1996). In Arabidopsis, there are seven genes encoding the same YIYTQ penta-peptide, with the exception of PSK6, which differs in the last amino acid (Figure 4).

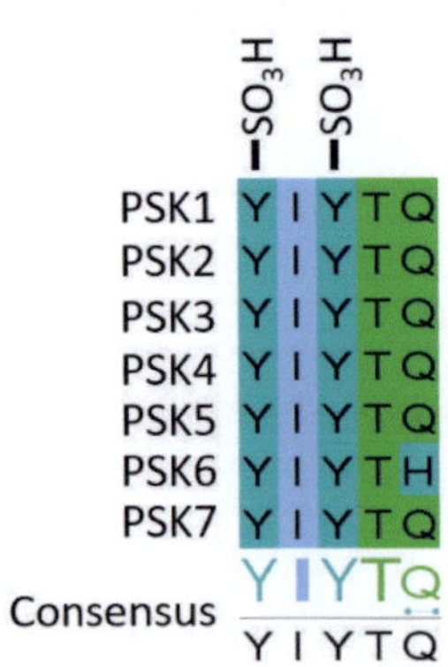

Figure 4. Alignment of the mature PSK peptides in Arabidopsis. Protein sequence alignment of the mature PSK peptides. Posttranslational modifications are indicated by SO_3H (tyrosine sulfation). Protein sequence alignment was done using MUSCLE (Multiple Sequence comparison by Log- expectation). The default Clustal X color scheme was used to assign colors to conserved amino acids based on their amino acid properties. PSK sequences were obtained from Kaufmann et al., 2019.

The maturation of PSK involves several steps. It may require pre-processing by SBT1.1, followed by proteolytic cleavage by unknown subtilases, and undergoes tyrosine sulfation by TPST (Srivastava et al., 2008). The proteolytic processing of PSK is unique because it involves the removal of the aspartate residue from the conserved DY motif at the N-terminus of sulfated peptides. The specific subtilase responsible for cleaving PSK to its final length in Arabidopsis is still unknown, while in tomato, it is catalyzed by phytaspase2 (Reichardt et al., 2020). In tomato, stress-induced PSK formation leads to flower drop (Reichardt et al., 2020). Further studies have revealed that PSK optimizes plant growth and defense by triggering the phosphorylation of the glutamine synthetase GS2 in tomato (Ding et al., 2022). In Arabidopsis, PSK has been attributed with various growth-promoting functions, including regulation of root growth, hypocotyl length, cell expansion, and pollen tube growth (Kutschmar et al., 2009; Stührwohldt et al., 2011; Hartmann et al., 2014; Stührwohldt et al., 2015). It also plays a role in plant immunity against diverse pathogens (Loivamäki et al., 2010; Igarashi et al., 2012; Mosher et al., 2012; Rodiuc et al., 2015). PSK functions through the LRR-RLKs PSKR1 and PSKR2, involving the co-receptor SERK3 (Ladwig et al., 2015; Hartmann et al., 2014; Kaufmann et al., 2021). Downstream signaling of the PSK/PSKR module involves CYCLIC NUCLEOTIDE-GATED CHANNEL17 and H^+-ATPase (Ladwig et al., 2015). Moreover, studies have revealed a potential application for PSK in the floral industry. Exogenous PSK treatment has been shown to delay the senescence of cut rose flowers and strawberry fruits (Aghdam et al., 2021a; Aghdam et al., 2021b). In summary, PSK has emerged as an important regulator of plant growth and development, as well as a mediator of stress responses in diverse plant species.

1.4.2 RGFs regulate responses beyond those specific to roots

The family of RGF (ROOT MERISTEM GROWTH FACTOR) peptides was initially identified by three independent research groups, resulting in three different names: CLAVATA 3/EMBRYO SURROUNDING REGION-LIKE (CLEL), ROOT MERISTEM GROWTH FACTOR (RGF), and GOLVEN (GLV) (Whitford et al., 2012; Meng et al., 2012; Matsuzaki et al., 2010). In Arabidopsis, the number of CLEL/RGF/GLV peptides varies depending on the study, with approximately 11 to 12 members. These peptides have a final length of 13-16 amino acids and contain a conserved

DY motif at the N-terminus, a conserved P (proline) fourth to the last amino acid, and a conserved N (asparagine) at the C-terminus (Figure 5).

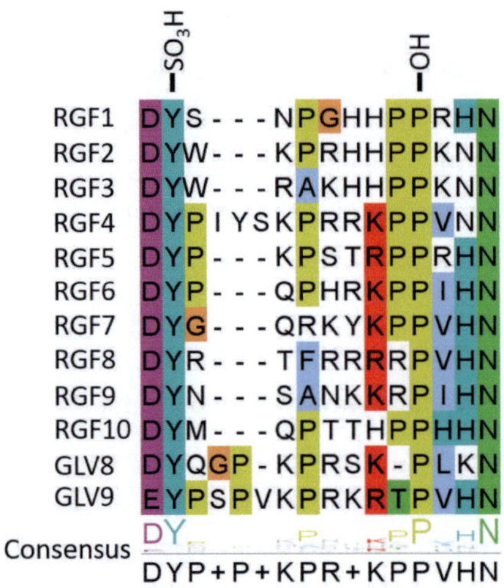

Figure 5. Alignment of the mature RGF peptides in Arabidopsis. Protein sequence alignment of the mature RGF/GLV peptides. Posttranslational modifications are indicated by SO_3H (tyrosine sulfation) and proline hydroxylation (OH). Protein sequence alignment was done using MUSCLE (Multiple Sequence comparison by Log- expectation). The default Clustal X color scheme was used to assign colors to conserved amino acids based on their amino acid properties. RGF/GLV sequences were obtained from Kaufmann et al., 2019.

The maturation process of RGF peptides involves tyrosine sulfation by TPST, proline hydroxylation at the conserved P residue, and proteolytic cleavage by subtilases. The aspartate residue at the N-terminus was found to be important for subtilase cleavage by SBT3.8 in the case of RGF1, RGF6, and RGF9, occurring in a post-Golgi compartment (Stührwohldt et al., 2020b; Stührwohldt et al., 2021). However, mutation of *sbt3.8* did not result in any specific phenotype related to RGF peptides, suggesting redundancy in N-terminal cleavage (Stührwohldt et al., 2020b). For RGF6 and RGF9, a pre-processing event by SBT6.1 at the RRLR/RRAL motifs of the precursor was shown to be essential for the function of these peptides (Ghorbani et al., 2016; Stührwohldt et al., 2020b). Most of the RGF peptides are primarily expressed in the roots, while RGF6 and RGF9 are primarily expressed in the aerial parts of the plant (Figure 6).

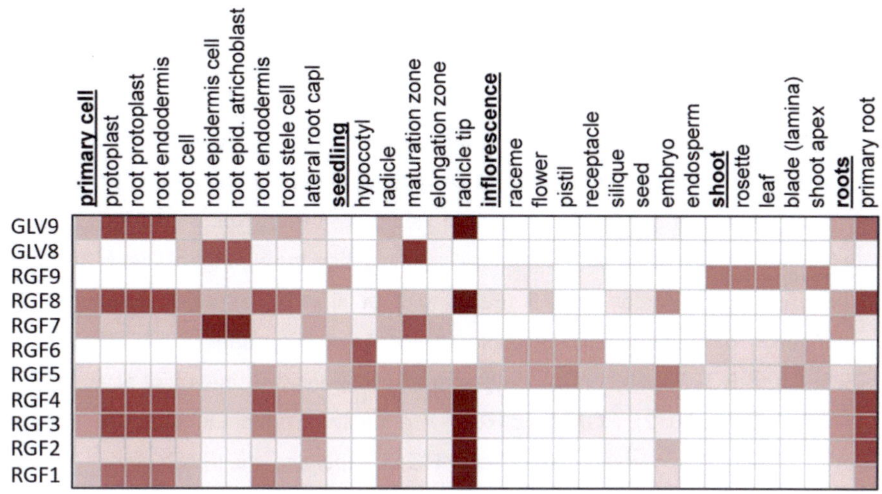

Figure 6. Expression-profiles of the RGF/GLV peptides. Tissue-specific expression profile of the RGF/GLV peptides in Arabidopsis. Colours indicate the rate of expression (red (high) and white (low)). The data set utilized for analysis was AT_mRNASeq_ARABI_GL-1, obtained through Genevestigator software. Adapted from Stegmann et al., 2022.

Several *rgf* single mutants display root-specific phenotypes like curly roots, increased root apical meristem size, and low lateral root density, indicating important functions in root development (Fernandez et al., 2013). In the search for an RGF1 receptor, the RGF1 INSENSITIVE (RGI) receptor family, consisting of five members, was identified (Meng et al., 2016; Song et al., 2016). RGF1 is redundantly perceived by the RGI1, RGI2, and RGI3 receptors (Jeon et al., 2022). The RGF1-RGI1 module activates a downstream YDA-MKK4/MKK5-MPK3/MPK6 signaling cascade, leading to altered expression levels of the PLETHORA (PLT1/PLT2) transcription factors, ultimately regulating the root stem cell niche maintenance (Lu et al., 2020; Shao et al., 2020; Fernandez et al., 2020a). Additionally, the RGI-RGF1 module controls the reactive oxygen species (ROS) accumulation in the developmental zones of the root through the RGF1-INDUCIBLE TRANSCRIPTION FACTOR 1 (RITF1) transcription factor, thus influencing the stability of PLT2 (Yamada et al., 2020; Jeon et al., 2021). Furthermore, RGI-RGF1 regulates lateral root development through MPK6 and the transcription factor PUCHI (Jeon et al., 2023). The SERK receptor family has been identified as essential co-receptors for RGI-RGF1-induced signaling (Ou et al., 2022; Jeon et al., 2023). Additionally, the two ubiquitin-specific proteases UBP12 and UBP13 regulate RGF1-RGI signaling by stabilizing the RGFR1 receptor (An et al., 2018). Likewise, the double mutant *ubp12/ubp13* is insensitive to RGF1 (An et al., 2018). Recently, RGF1 was identified as an extracellular pH sensor, as the pattern-triggered immunity (PTI)-induced alkalinization of the acidic extracellular pH in the apical meristem inhibits the tyrosine sulfate-dependent interaction of RGF1 with RGI1 (Liu et al., 2022). Similar to RGF1 signaling, RGF5 and RGF8 inhibit asymmetric cell divisions during lateral root initiation through the RGIs, MPK6, and the transcription factor PUCHI (Fernandez et al., 2020b). RGF6 regulates the elongation of the hypocotyl of etiolated seedlings (Ghorbani et al., 2016), while RGF6 and RGF9 both control the gravitropic response (Stührwohldt et al., 2020b; Whitford et al., 2012). RGF9 and RGF7 are involved in plant immunity (Stegmann et al., 2022; Wang et al., 2020), and RGF4 regulates the PIN2 amount in root epidermal cells, thereby interfering with the auxin-induced gravitropic response (Whitford et al., 2012). The full scope of biological functions of many RGF peptides is still to be discovered.

1.4.3 PSY peptides balance stress and growth responses

The family of PSY peptides consists of nine members in Arabidopsis (Ogawa-Ohnishi et al., 2022). The first identified member, PSY1, has a length of 18 amino acids, contains a tyrosine sulfation and two hydroxyproline residues, of which one carries an O-linked glycosylation (Amano et al., 2007). PSY1R functions as the PSY1 receptor and is related to the PSK receptors (Amano et al., 2007). The PSY1-PSY1R module controls cell size, root growth, the formation of the seedling cuticle, and activates the plasma membrane proton pump AHA2/AHA1 by phosphorylation (Amano et al., 2007; Pruitt et al., 2017; Giorgi et al., 2021; Fuglsang et al., 2014).

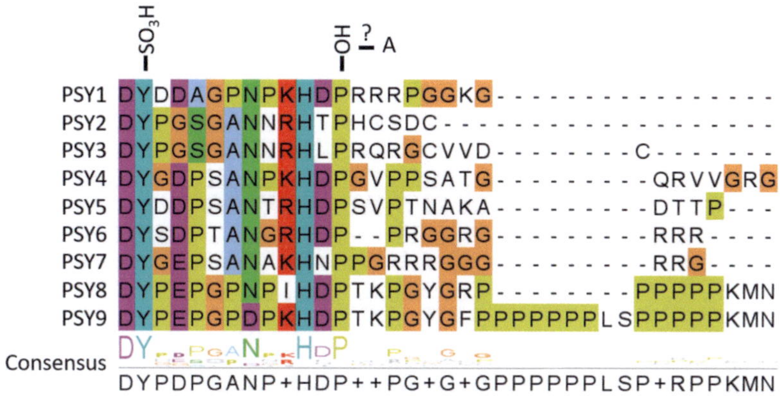

Figure 7. Alignment of the mature PSY peptides in Arabidopsis. Protein sequence alignment of the mature PSY peptides. Posttranslational modifications are indicated by SO_3H (tyrosine sulfation) and OH (proline hydroxylation) with potential O-glycosylation (?A). Protein sequence alignment was done using MUSCLE (Multiple Sequence comparison by Log- expectation). The default Clustal X color scheme was used to assign colors to conserved amino acids based on their amino acid properties. Since not all mature PSY sequences are yet discovered, the complete C-term is illustrated. PSY sequences were obtained from Ogawa-Ohnishi et al., 2022.

PSY1, PSY3, PSY5, PSY6, and PSY8 are expressed all over the plant in the inflorescence, shoots, and roots (Figure 8). PSY7 expression is limited to the inflorescence, PSY2 and PSY5 to shoots and inflorescence, while PSY4 expression is only detectable in the roots (Figure 8). Thus, indicating various functions for PSYs all over the plant. A recent study proposed that the PSY peptides regulate abiotic stress responses in Arabidopsis, as many PSY transcripts are upregulated upon abscisic acid treatment (Tost et al., 2021).

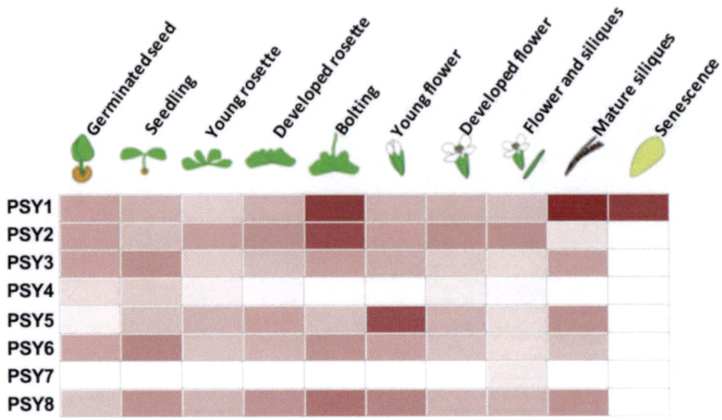

Figure 8. Expression profiles of PSY peptides in Arabidopsis. Expression levels of PSY1-8 were determined using mRNAseq data sourced from Genevestigator. Colours indicate the rate of expression (red (high) and white (low)). Adapted from Tost et al., 2021.

The biotrophic pathogen Xanthomonas oryzae produces a sulfated peptide that mimics the PSYs to increase its virulence to plants, thereby exploiting the important functions of PSY peptides (Pruitt et al., 2017). In addition to the PSY1/PSY1R module, three receptors (PSYR1, PSYR2, and PSYR3) with a different PSY signaling were reported in a subsequent study (Ogawa-Ohnishi et al., 2022). The study identified the mature sequences of PSY5, PSY6, and PSY8, which, in contrast to PSY1, did not show hydroxyproline arabinosylation (Ogawa-Ohnishi et al., 2022). A generated *psy5/psy6/psy8* triple mutant does not exhibit any growth defects, indicating redundancy among the PSY peptides (Ogawa-Ohnishi et al., 2022). All nine PSY peptides could interact with the PSYR1, PSYR2, and PSYR3 receptors, consistent with the proposed redundancy (Ogawa-Ohnishi et al., 2022). By utilizing a *psyr1/psyr2/psyr3* triple mutant, the PSYRs were found to induce stress-responsive genes in the absence of their ligands, ultimately controlling the trade-off between plant growth and stress response (Ogawa-Ohnishi et al., 2022). In contrast to prior studies, the first identified PSY1R for PSY1 was not relevant in this signaling pathway. Additionally, receptors from the group of SERKs are not relevant in this mechanism, suggesting the involvement of unknown co-receptors in PSYR signaling.

1.4.4 CIF and TWS1 regulate formation of barriers

The family of the CIFs consists of five members that contain the DY motif, two conserved proline residues that are hydroxylated (only shown in CIF1), a conserved hydrophobic amino acid (I or V), and an N or H amino acid at the end of the sequence (Figure 8). They all function through the GSO1 and GSO2 receptors (Nakayama et al., 2017; Doblas et al., 2017; Doll et al., 2020; Truskina et al., 2022). CIF1 and CIF2 are required for the Casparian strip diffusion barrier formation and lignification in Arabidopsis roots (Nakayama et al., 2017; Doblas et al., 2017). CIF1 and CIF2 are expressed in the stele (vascular tissue) and travel to their receptor GSO1, which is specifically expressed at the outer, cortex-facing side of the endodermal plasma membrane (Doblas et al., 2017). The CIF peptides induce the formation of the Casparian strip, which forms a barrier between the stele and cortex, preventing the CIFs from reaching their receptor and thereby self-regulating the termination of this mechanism.

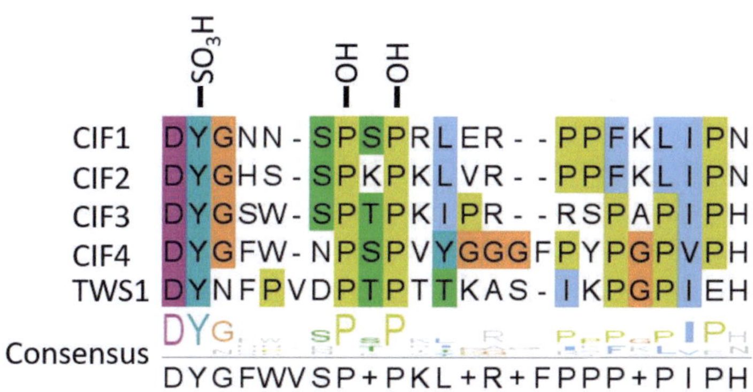

Figure 9. Alignment of the mature CIF peptides in Arabidopsis. Protein sequence alignment of the mature PSY peptides. Posttranslational modifications are indicated by SO_3H (tyrosine sulfation) and OH (proline hydroxylation). Protein sequence alignment was done using MUSCLE (Multiple Sequence comparison by Log- expectation). The default Clustal X color scheme was used to assign colors to conserved amino acids based on their amino acid properties. CIF/TWS sequences were obtained from Truskina et al., 2022.

A further study revealed an additional function for CIF2, that together with PSY1, is sulfated in the endosperm and subsequently released to bind to the PSY1R or

GSO1/GSO2 receptor on the embryo to induce cuticle formation (Giorgi et al., 2021). Grafting experiments showed that the endosperm is required for a well-sealed cuticle. The lack of endosperm, as well as the lack of sulfation of the endosperm-released PSY1 and CIF2, leads to defects in root development and chloroplast functions (Giorgi et al., 2021). The phenotypes displayed by the *psy1/cif1/cif2* triple mutant are weaker compared to those of the *tpst*-mutant, indicating the involvement of additional sulfated peptides in this process (Giorgi et al., 2021).

Another member of the CIF family, TWISTED SEED1 (TWS1), is required for cuticle formation. The TWS1 precursor is synthesized and tyrosine-sulfated in the embryo and released into the endosperm where it undergoes proteolytic cleavage by SBT2.4 (ALE1) and SBT1.8 (Doll et al., 2020; Royek et al., 2022). The cleavage of the TWS1 precursor by SBT1.8 is dependent on tyrosine sulfation (Royek et al., 2022). Subsequently, mature TWS1 travels back to the embryo to induce the formation of the embryonic cuticle through the GSO1/GSO2 receptors using SERK1, SERK2, and SERK3 as co-receptors (Doll et al., 2020; Okuda et al., 2020; Zhang et al., 2022). Similar to the mechanism of CIF1/CIF2 Casparian strip formation, the induction by TWS1 will last until the cuticle is formed and TWS1 cannot pass through (Doblas et al., 2017; Fujita et al., 2021). CIF3 and CIF4 have an important role in pollen wall development. They are expressed in the tapetum cells of developing anthers, and their activation is dependent on the cleavage by SBT5.4, which is produced by developing pollen grains (Truskina et al., 2022). Mature CIF3 and CIF4 peptides bind their receptors (GSO1/GSO2) in the middle layer, inducing pollen wall development (Truskina et al., 2022).

1.5 Aim of the dissertation

Recently, sulfated peptides have emerged as important regulators in plant growth, development, and stress responses. Thus, the focus of this dissertation lies on the analysis of the functions of sulfated peptides in Arabidopsis. These peptides are synthesized as larger precursors and undergo posttranslational modifications, such as tyrosine sulfation by TPST, proteolytic cleavage by subtilases, proline hydroxylation, or O-linked glycosylation. For instance, tyrosine sulfation is a crucial modification for receptor binding and subsequent posttranslational modification events, underlining the importance of this modification. To trigger downstream signaling, mature peptides are perceived by receptors from the LRR-RLK family. Although specific sulfated peptides have been extensively studied and proposed to have important roles in plant development, many remain without a known function. Likewise, there is a lack of information about posttranslational modification events of sulfated peptides, especially cleavage events involving subtilases and perceiving receptors. Thus, the objective of this work focuses on the maturation process, receptor binding, and the biological functions of sulfated peptides. The *tpst* mutant, which can be considered a loss-of-function mutant for all sulfated peptides, is utilized to analyze the role of sulfated peptides in Arabidopsis. Due to the pleiotropic phenotypes associated with the *tpst* mutant, the mature sulfated peptides are used to complement the *tpst* phenotypes. The main focus lies on investigating the functions of PSK and RGF peptides.

PSK has been attributed significant functions in various species, including triggering drought-induced flower drop in tomato, suggesting a potential role in drought resistance also in Arabidopsis. While the subtilases responsible for cleaving the PSK precursor in tomato, known as phytaspase2, have been identified, the corresponding subtilase in Arabidopsis remains unknown. This study aims to investigate PSK maturation in Arabidopsis and its role in drought resistance.

In contrast to most RGFs, RGF6 and RGF9 are primarily expressed in the hypocotyl and leaves of the plant. Initially identified as root growth factors, addressing the distinct expression pattern of RGF6 and RGF9, this research investigates their potential functions in non-root tissues. Since the proteolytic processing of SBT6.1 and the tyrosine sulfation of TPST are essential for the peptides function, loss of function mutant of those should be used as tools to analyse the peptides function.

2. Results

2.1 Sulfopeptide CLEL6 inhibits anthocyanin biosynthesis in *Arabidopsis thaliana*

Running title: CLEL6-regulated anthocyanin biosynthesis

Eric Bühler[1], Elisa Fahrbach[1], Andreas Schaller[1] and Nils Stührwohldt[1,#]

[1] Department of Plant Physiology and Biochemistry, Institute of Biology, University of Hohenheim, 70593 Stuttgart, Germany.

[#]correspondence: nils.stuehrwohldt@uni-hohenheim.de

Status: Published

Original publication: Plant Physiology, kiad316, https://doi.org/10.1093/plphys/kiad316

Publication available at: https://academic.oup.com/plphys/advance-article-abstract/doi/10.1093/plphys/kiad316/7186972?redirectedFrom=fulltext

Own contribution: I contributed to the study design, performed all experiments and wrote the manuscript together with Nils Stührwohldt and Andreas Schaller.

Abstract

Post-translationally modified peptides are now recognized as important regulators of plant stress responses. Here we identified the small sulfated CLE-LIKE6 (CLEL6) peptide as a negative regulator of anthocyanin biosynthesis in etiolated and in light-stressed Arabidopsis seedlings. The function of CLEL6 depends on proteolytic processing of the CLEL6 precursor by the subtilisin-like serine proteinase 6.1 (SBT6.1), and on tyrosine sulfation by tyrosylprotein sulfotransferase (TPST). Loss of function mutants of either *sbt6.1* or *tpst* showed significantly higher anthocyanin accumulation upon light stress. The overaccumulation phenotype of *sbt6.1* and *tpst* was suppressed by application of mature CLEL6. Further confirming the role of CLEL6 as an inhibitor of anthocyanin biosynthesis, overexpression and external application of CLEL6 inhibited the expression of genes involved in anthocyanin biosynthesis in etiolated and light- stressed seedlings. Small post-translationally modified peptides are known to be perceived by leucine-rich-repeat receptor like kinases. Through a genetic approach, using a ROOT MERISTEM GROWTH FACTOR 1 INSENSITIVE (RGI) receptor quintuple mutant, we could show the essential function of the RGI receptor family in CLEL6 signaling. Our data indicate that overexpresssion or application of CLEL6 inhibit anthocyanin biosynthesis through RGI receptors. We propose that CLEL6 inhibits anthocyanin biosynthesis in etiolated seedlings, and that the suppression of anthocyanin biosynthesis is released when CLEL6 expression is down-regulated upon light exposure. Hyperaccumulation of anthocyanins in light-stressed *tpst* and *sbt6.1* mutant seedlings suggests that CLEL6, or related sulfopeptides, continue to act as neagtive regulators to limit pigment accumulation in the light.

Introduction

The apical hook, hypocotyl elongation, and closed cotyledons are hallmarks of skotomorphogenic development of dark-grown seedlings struggling to reach the light. Skotomorphogenesis relies on a group of basic helix-loop-helix transcriptional repressors named PHYTOCHROME-INTERACTING FACTORs (PIFs) that inhibit the expression photomorphogenic genes like *HEMA1* for glutamyl-$tRNA^{Glu}$ reductase, *GENOMES UNCOUPLED5* (*GUN5*), *PHOTOSYSTEM I LIGHT HARVESTING COMPLEX GENE1* (*LHCA1*) and *PS-AE1* for Photosystem I reaction center subunit IV A (Duek and Fankhauser, 2005; Monte et al., 2007; Shin et al., 2009). In addition, positive regulators of photomorphogenesis, the bZIP transcription factors ELONGATED HYPOCOTYL 5 (HY5) and LONG HYPOCOTYL IN FAR-RED (HFR1), are targeted for degradation by the ubiquitin ligase complex CONSTITUTIVE PHOTOMORPHOGENESIS 1-SUPPRESSOR OF PHYA-105 (COP1-SPA) in the dark (Osterlund et al., 2000; Jang et al., 2005). Upon perception of light, phytochrome photoreceptors are activated leading to the degradation of PIFs and re-organization of the COP1-SPA complex. Thereby, photomorphogenic genes are induced, stem elongation comes to a halt, the apical hook opens, cotyledons expand, functional chloroplasts are formed, and photosynthesis ensues (Leivar et al., 2008; Jiao et al., 2007; Sheerin et al., 2015).

When chloroplast biogenesis is disturbed, or when chloroplasts are dysfunctional as a result of environmental stress, GENOMES UNCOUPLED (GUN1) mediates retrograde plastid-to-nucleus signaling to inhibit the expression of photomorphogenic genes and to prevent photo-oxidative damage (Koussevitzky et al., 2007). GUN1-mediated retrograde signaling also controls the accumulation of anti-oxidative anthocyanins that protect plants from stress-induced reactive oxygen species (ROS) (Richter et al., 2020; Nakabayashi et al., 2014; Shao et al., 2007). In addition to their anti-oxidative activity, anthocyanins also absorb light directly, thereby reducing excitation pressure under high irradiance (Steyn et al., 2002; Xu et al., 2017). Consequently, anthocyanins alleviate light stress during de-etiolation and early seedling establishment, while the photosynthetic apparatus is not yet fully assembled, and later they protect the photosynthetic apparatus from photo inhibition (Steyn et al., 2002; Drumm-Herrel and Mohr, 1985). Consistent with their relevance for light stress protection, the formation of anthocyanins is suppressed in dark-grown seedlings (Shin et al., 2007). Activation of the genes in the anthocyanin biosynthetic pathway depends on the general regulator of photomorphogenesis COP1 (Deng et al., 1991). Anthocyanin biosynthesis genes are grouped into early genes including *CHALCONE SYNTHASE* (*CHS*), *CHALCONE ISOMERASE* (*CHI*), and *FLAVANONE 3-HYDROXYLASE* (*F3H*), and late genes including *DIHYDROFLAVONOL 4-REDUCTASE* (*DFR*), *LEUCOANTHOCYANIDIN OXYGENASE* (*LDOX*) and *UDP-GLUCOSE:FLAVONOID 3- O-GLUCOSYLTRANSFERASE* (*UF3GT*). Expression of these genes is controlled by numerous R2R3 MYB transcription factors, including MYB11, MYB12 and MYB111 for the regulation of early biosynthesis genes (Stracke et al., 2007), and PRODUCTION OF ANTHOCYANIN PIGMENT1 (PAP1) for late biosynthesis genes (Xie et al., 2006). The GOLDEN-2-LIKE1 (GLK1) transcription factor mediates anthocyanin accumulation in response to sucrose (Zhao et al., 2021).

We report here that peptide signaling contributes to the regulation of anthocyanin biosynthesis in dark-grown as well as in de-etiolated and light-stressed seedlings. Post-translationally modified peptides, particularly those containing a sulfated tyrosine residue, have previously been implicated in stress response signaling. For example, PHYTOSULFOKINE (PSK) contributes to drought stress tolerance in Arabidopsis (Stührwohldt et al., 2021). In tomato, PSK triggers drought-induced abscission of flowers and fruits (Reichardt et al., 2020). Immune responses in Arabidopsis and tomato are also regulated by PSK signaling (Mosher et al., 2013; Zhang et al., 2018; Igarashi et al., 2012). Likewise, CLE-LIKE9 (CLEL9) signaling enhances resistance against bacterial pathogens, and CLEL4 triggers innate immune responses in Arabidopsis (Stegmann et al., 2022; Wang et al., 2021). PLANT PEPTIDE CONTAINING SULFATED TYROSINE (PSY) peptides have been implicated in abscisic acid-mediated stress responses and immune signaling (Tost et al., 2021), and unidentified sulfated peptides are involved in the response to phosphate deficiency (Kang et al., 2014). However, a role for peptides in light stress signaling, the regulation of anthocyanin production, or photomorphogenesis has not been reported so far. The formation of sulfated signaling peptides relies on subtilisin-like serine proteinases (subtilases, SBTs) for proteolytic processing (Schaller et al., 2018), and on tyrosylprotein sulfotransferase (TPST) for post-translational sulfation (Komori et al., 2009). In case of CLEL peptides, processing by SBT6.1 is required for maturation and secretion, and sulfation by TPST for bioactivity (Stührwohldt et al., 2020b; Ghorbani et al., 2016). Bioactivity of PSK depends on processing by SBT3.8 and Phyt2 in Arabidopsis and tomato, respectively (Stührwohldt et al., 2021; Reichardt et al., 2020), and on TPST for receptor recognition (Wang et al., 2015). In case of TWISTED SEED1 (TWS1), tyrosine sulfation is required for both, receptor binding and processing by SBT1.8 (Royek et al., 2022; Doll et al., 2020; Stintzi and Schaller, 2022).

In this work we show that the sulfopeptide CLEL6 is a negative regulator of anthocyanin accumulation in Arabidopsis, and that the regulatory function of CLEL6 depends on proteolytic processing by SBT6.1, tyrosine sulfation by TPST, and perception by RGF1 INSENSITIVE (RGI) receptors. We further show that CLEL6-mediated suppression of anthocyanin biosynthesis in etiolated seedlings is released when *CLEL6* expression is down-regulated upon transition to light, allowing for anthocyanin biosynthesis and fine- tuning of pigment production.

Results

Photomorphogenesis and/or anthocyanin biosynthesis are mis-regulated in *tpst-1* and *sbt6.1*

TPST is the only tyrosylprotein sulfotransferase in Arabidopsis and is essential for post- translational sulfation of tyrosine residues (Komori et al., 2009). Loss of TPST function thus results in pleiotropic defects that can be attributed to the impaired production of any one of the many sulfated signaling peptides (Kaufmann and Sauter, 2019). We observed that etiolated *tpst-1* seedlings display short hypocotyls and open cotyledons (Figure 1), resembling the constitutive photomorphogenic phenotype of the *cop1* mutant (Deng et al., 1991). We conclude that TPST activity is required to suppress photomorphogenesis in the dark.

During skotomorphogenesis, the COP1-SPA ubiquitin ligase complex mediates the degradation of PAP1 and PAP2 transcription factors to prevent the formation of anthocyanins (Maier et al., 2013). Consequently, in the *cop1* mutant, PAP transcription factors are stabilized resulting in the expression of anthocyanin biosynthetic genes and the accumulation of anthocyanins in darkness (Deng et al., 1991). Similarly, we observed strong (up to 50-fold) upregulation of early (*CHS*, *F3H*) and late (*LDOX*, *UF3GT*, *UGT78D2*) anthocyanin biosynthesis genes, and the accumulation of anthocyanins in etiolated *tpst-1* compared to wild-type seedlings (Figure 2, B and C). Light-independent accumulation of anthocyanins and upregulation of anthocyanin- related genes was confirmed in an independent *tpst* allele (*tpst-2*, aka *hps7*; (Kang et al., 2014) (Supplemental Figure S1, A and B).

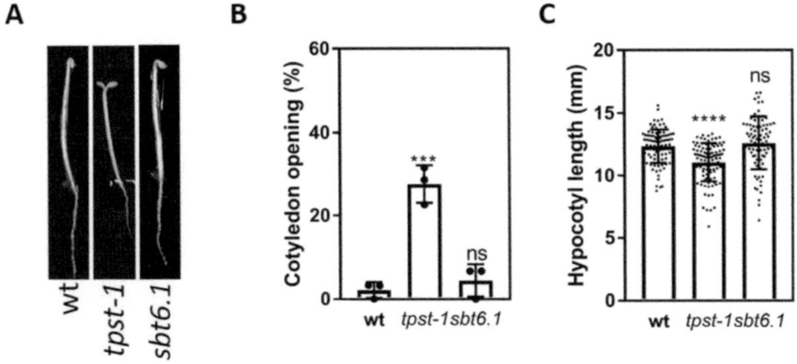

Figure 1: *tpst-1* displays an early photomorphogenesis phenotype. A Representative five-day-old etiolated wild-type (wt), *tpst-1* and *sbt6.1* seedlings. **B, C** Quantification of **(B)** open cotyledons (in %) and **(C)** hypocotyl length (mm) of five- day-old etiolated wt, *tpst-1* and *sbt6.1* seedlings. Data are shown as the mean ± SD. **(B)** n=3 independent experiment with at least 29 seedlings each. **(C)** n=3 independent experiment with at least 26 seedlings each. Asterisks indicate significant differences (***P<0.001, ****P<0.0001; two-tailed t-test).

Interestingly, loss of TPST function caused enhanced anthocyanin production also in response to light stress. When five-day-old etiolated seedlings were exposed to light for 16 hours, the accumulation of anthocyanins was higher in *tpst-1* compared to the wild type (Figure 2, A and B), and it correlated with increased expression of many of the anthocyanin biosynthesis genes two hours after light exposure (Figure 2E). The data indicate that TPST activity suppresses anthocyanin formation in dark-grown seedlings and limits the accumulation of anthocyanins in the light. Considering its essential role in the biogenesis of signaling peptides, we hypothesize that constitutive photomorphogenesis and enhanced anthocyanin production in the TPST loss of function mutant are caused by the lack of one or more sulfopeptides that act as negative regulator(s) of these processes.

To corroborate this notion, and to narrow down the number of sulfopeptide candidates, we analyzed *sbt6.1* as a second mutant that also is impaired in the production of signaling peptides. In addition to various other proteins (Schaller et al., 2018), SBT6.1 is known to be involved in the maturation of RALFs among the cysteine-rich signaling peptides (Stegmann et al., 2017), and of CLELs among the post-translationally sulfated peptides (Stührwohldt et al., 2020b; Ghorbani et al., 2016). The constitutive photomorphogenic phenotype of the *tpst-1* mutant was not observed in *sbt6.1* (Figure 1). The cotyledons of *sbt6.1* seedlings did not open in darkness and, unlike (Ghorbani et al., 2016), we also did not detect any reduction in hypocotyl elongation in dark-grown *sbt6.1* compared to wild-type seedlings (Figure 1).

In contrast, mis-regulation of anthocyanin biosynthesis was also observed in *sbt6.1*. The same anthocyanin biosynthesis genes that were upregulated in dark-grown *tpst-1* compared to wild-type seedlings showed increased expression also in *sbt6.1* (Figure 2D). The level of induction was lower in *sbt6.1* compared to *tpst-1* (Figure 2, C and D), and apparently not sufficient to result in a significant accumulation of anthocyanins in darkness (Figure 2B). In response to light-stress, on the other hand, expression levels of anthocyanin biosynthesis genes in mutant vs. wild-type seedlings were increased to a similar extent in *tpst-1* and *sbt6.1* (Figure 2, E and F), resulting in a similar increase in anthocyanin production compared to the wild type (Figure 2B). Also, in continuous light, we observed substantially higher levels of anthocyanins in *tpst-1*, *tpst-2* and *sbt6.1* than in wild-type seedlings (Supplemental Figure S2). Our data suggest that photomorphogenesis (cotyledon opening and inhibition of hypocotyl elongation) and anthocyanin production are negatively regulated by sulfated signaling peptides. However, since the two processes are differentially affected in *tpst* and *sbt6.1* mutants, the peptides involved are not necessarily the same. In the following we focused on the peptide-mediated regulation of anthocyanin biosynthesis.

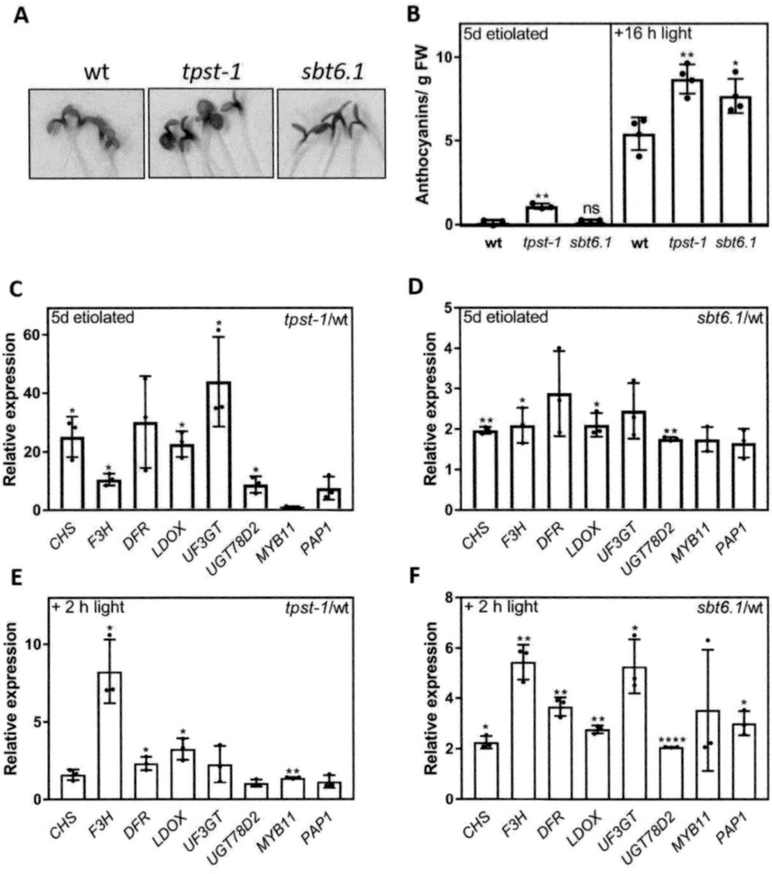

Figure 2: TPST and SBT6.1 signaling represses anthocyanin biosynthesis during light stress. **A** Representative wild-type (wt), *tpst-1* and *sbt6.1* seedlings grown for five-days in the dark followed by 16 h in the light. **B** Quantification of anthocyanin content in five-day-old etiolated wt, *tpst-1* and *sbt6.1* seedlings prior and after 16 h light exposure. **C, D, E, F** Relative expression of *CHS*, *F3H*, *DFR*, *LDOX*, *UF3GT*, *UGT78D2*, *PAP1*, *MYB11* in five-day-old etiolated *tpst-1* or *sbt6.1* seedlings compared to the wt control. In **E, F** the seedlings were exposed to light for 2 h before sampling. Gene expression was normalized to two reference genes. Data are shown as the mean ±SD for n=3 (B etiolated, C, D, E, F) or n=4 (B + 16 h light) independent experiments. Asterisks indicate significant differences (*P<0.05, **P<0.01, ****P<0.0001; (B) two-tailed *t*-test; (C, D, E, F) one sample *t*-test).

CLEL6 is a negative regulator of anthocyanin biosynthesis

The formation of anthocyanins is suppressed in etiolated seedlings in a TPST-dependent manner, and in response to light stress, anthocyanins accumulate to higher levels in *tpst-1* and *sbt6.1* than in wild-type seedlings (Figure 2B). We attribute the increase in anthocyanins in *tpst-1* and *sbt6.1* to the loss of a putative signaling peptide that acts as a negative regulator to restrict anthocyanins in the wild type. The abundance (or activity) of this putative negative regulator is expected to be high in dark-grown seedlings when anthocyanin synthesis is suppressed, and less so in light-stressed seedlings when anthocyanins are formed. Therefore, to identify this negative regulator, we looked for candidate peptides that are expressed at lower levels in light-treated than in etiolated seedlings. The expression of all hitherto known genes for sulfated signaling peptides (Kaufmann and Sauter, 2019) was analyzed, and CLEL6 was found to be the only one expressed at significantly (3.5-fold) lower level at two hours after light exposure compared to etiolated seedlings (Supplemental Figure S3A). This result was confirmed for seedlings grown in continuous light, that also showed lower CLEL6 expression than dark-grown seedlings (Supplemental Figure S3B). Consistent with a role for CLEL6 as a candidate regulator of anthocyanin biosynthesis are previous findings indicating that CLEL6 bioactivity relies on both, SBT6.1 for precursor processing, and TPST for tyrosine sulfation (Stührwohldt et al., 2020b; Ghorbani et al., 2016). Further, and in contrast to most other CLEL genes that are active predominantly in the root, CLEL6 and the closely related CLEL9 are shoot-specific and mainly expressed in the hypocotyl and leaves, which matches the anthocyanin accumulating tissues (Fernandez et al., 2013).

Unlike CLEL6, however, the expression of CLEL9 was not reduced in response to light (Supplemental Figure S3, A and C). The suggested involvement of CLEL6 in the regulation of anthocyanin biosynthesis was tested by addition of the mature CLEL6 peptide (DY(SO3H)PQPHRKPPIHNE) to the growth medium. Addition of CLEL6 reduced anthocyanin levels in light-stressed *tpst-1* and *sbt6.1* to that of the wild type (Figure 3, A and B). Also in *tpst-2*, the anthocyanin overaccumulation phenotype was complemented by treatment with CLEL6 (Supplemental Figure S4A). The inhibitory effect of CLEL6 on anthocyanin production was observed at concentrations as low as 10 nM, with highest activity at 1 µM, which was used in all other experiments (Supplemental Figure S5). Notably, unsulfated CLEL6 and the closely related CLEL9 peptide (DMDY(SO3H)NSANKKRPIHN) were inactive in this assay (Figure 3A and Supplemental Figure S4B).

To test CLEL6 as a possible suppressor of anthocyanin formation during skotomorphgenesis, we grew *tpst-1* seedlings on media supplemented with the mature CLEL6, unsulfated CLEL6, or CLEL9 peptides in darkness. In contrast to CLEL9 and unsulfated CLEL6 which had no significant effects (Figure 3C and Supplemental Figure S6), CLEL6 treatment led to a substantial reduction in anthocyanin content of etiolated *tpst-1* seedlings (Figure 3C). Consistently, the expression level of most of the anthocyanin biosynthesis genes was much reduced by CLEL6 treatment (Figure 3D). Finally, we tested the effect of CLEL6 on anthocyanin formation in continuous light. Wild- type, *tpst-1*, *tpst-2* and *sbt6.1* seedlings grown on CLEL6-supplemented media accumulated reduced amounts of anthocyanins compared to plants on control plates (Supplemental Figure S7). Taken together, these data suggest a specific role for CLEL6 as the negative regulator of anthocyanin biosynthesis in the light and in darkness as well.

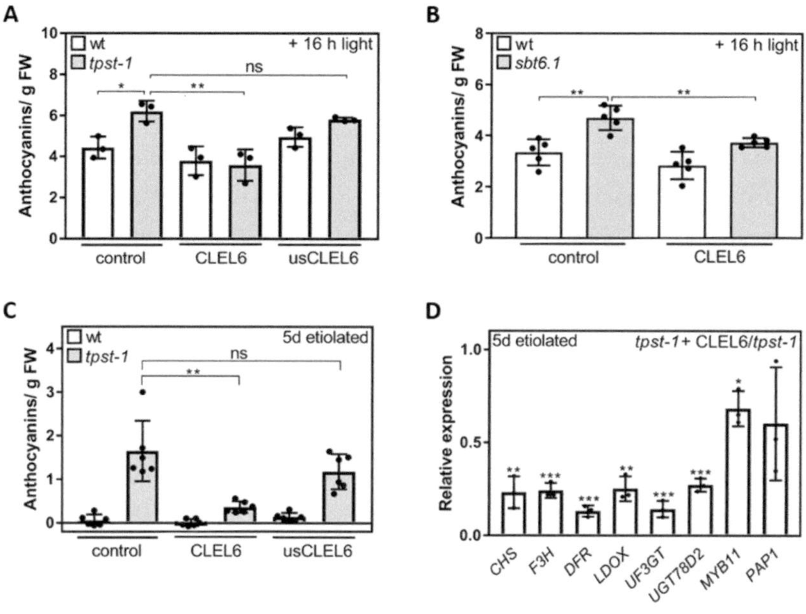

Figure 3: CLEL6 suppresses anthocyanin accumulation and gene expression. A Quantification of anthocyanin content of five-day-old etiolated wt and *tpst-1* seedlings after 16 h light exposure. 1 µM CLEL6 or 1 µM usCLEL6 were added to the media as indicated. **B** Quantification of anthocyanin content of five-day-old etiolated wt and *sbt6.1* seedlings after 16 h light exposure. 1 µM CLEL6 was added to the media as indicated. **C** Quantification of the anthocyanin content in five-day-old etiolated wt and *tpst-1* seedlings with 1 µM CLEL6 or 1 µM unsulfated CLEL6 (usCLEL6) added in the media as indicated. **D** Relative Expression of *CHS*, *F3H*, *DFR*, *LDOX*, *UF3GT*, *UGT78D2*, *PAP1* and *MYB11* of five-day-old etiolated *tpst-1* seedlings grown on a media with 1 µM CLEL6 relative to the *tpst-1* control. Gene expression was normalized to two reference genes. Data are shown as the mean ±SD of x n=3 (A, D) or n=5 (B) or n=6 (C) independent experiments. Asterisks indicate significant differences (**P*<0.05, ***P*<0.01, ****P*<0.001; (A, B, C) two-tailed *t*- test; (D) one sample *t*-test).

Overexpression of CLEL6 inhibits anthocyanin production

Consistent with the proposed role for CLEL6 as a negative regulator of anthocyanin biosynthesis, we observed that anthocyanin accumulation was impaired in transgenic plants constitutively overexpressing CLEL6. *35S:CLEL6* seedlings produced less anthocyanins than the wild type in response to light stress (Figure 4, A and B) and under continuous light (Supplemental Figure S8A), and the expression of most of the anthocyanin biosynthesis genes was strongly reduced in light-stressed *35S:CLEL6* compared to wild-type seedlings (Figure 4D). In contrast, in transgenic *35S:CLEL9* seedlings, the overexpression of CLEL9 did not affect anthocyanin biosynthesis and accumulation, confirming the specific role of CLEL6 in this process (Supplemental Figure S8B). Also in dark-grown seedlings, the expression of anthocyanin biosynthesis genes was much lower in *35S:CLEL6* transgenics compared to the wild type, and formation of anthocyanins was not observed (Figure 4, B and C), while overexpression of CLEL9 (*35S:CLEL9*) had no effect on anthocyanin content (Supplemental Figure S8B).

The results are consistent with those obtained in peptide complementation experiments (Figure 3 A-D, Supplemental Figures S4A, S5 and S7) and confirm a specific role for CLEL6 in the downregulation of anthocyanin biosynthesis in dark and light whereas the closest homolog, CLEL9, does not seem to be involved.

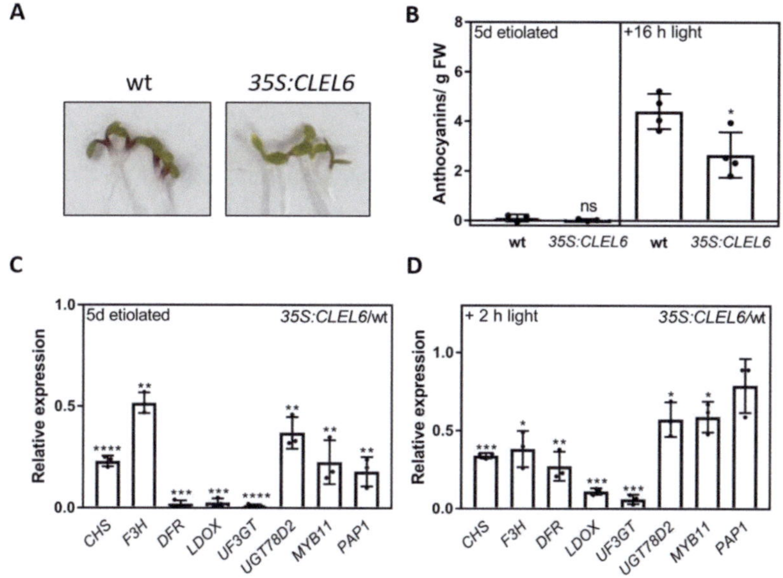

Figure 4: Overexpression of CLEL6 suppresses anthocyanin accumulation and biosynthesis genes. A Representative wt and *35S:CLEL6* seedlings grown for five- days in the dark followed by 16 h in the light. **B** Quantification of anthocyanin content in five-day-old etiolated wt and *35S:CLEL6* seedlings prior and after 16 h light exposure. **C, D** Relative Expression of *CHS*, *F3H*, *DFR*, *LDOX*, *UF3GT*, *UGT78D2*, *PAP1*, *MYB11* in five-day-old etiolated *35S:CLEL6* seedlings compared to the wt control. In **D** seedlings were exposed to light for 2 h before sampling. Gene expression was normalized to two reference genes. Data are shown as the mean ±SD of B n=4 (B + 16 h light) or n=3 (B etiolated, C, D) independent experiments. Asterisks indicate significant differences (*P<0.05, **P<0.01, ***P<0.001; ****P<0.0001; (B) two-tailed t-test; (C, D) one sample t-test).

CLEL6 signaling depends on RGIs

Within the large group of leucine-rich repeat receptor kinases (LRR-RKs), the five-membered RGI (ROOT MERISTEM GROWTH FACTOR (RGF)1 INSENSITIVE) family comprises the receptors of CLEL peptides. While RGI1 and RGI2 are mostly expressed in the root tip (hence their name), RGI3-5 are also expressed in leaves (Song et al., 2016; Stegmann et al., 2022). RGI receptors act redundantly in the perception of CLEL8 (aka RGF1) for the regulation of root meristem development (Ou et al., 2016). RGI3 perceives CLEL9 (RGF9) and is required for its immune-modulatory activity (Stegmann et al., 2022). Likewise, the activity of CLEL4 (RGF7) as a damage-associated molecular pattern relies on the RGI4 and RGI5 receptors (Wang et al., 2021). To address the question whether the RGIs are important for CLEL6 signaling, we used RGI receptor mutants. In contrast to single gene loss-of-function mutants (Supplemental Figure S9A), the RGI quintuple mutant (*rgi5x*; (Ou et al., 2016) accumulated significantly more anthocyanin than the wild type after light exposure, thus resembling the CLEL6 loss-of- function phenotype of the *sbt6.1* and *tpst* mutants (Figure 5A). In darkness, *rgi5x* seedlings were more similar to *sbt6.1* than to *tpst-1*, as they showed neither increased anthocyanin accumulation (Supplemental Figure S9B) nor cotyledon opening, but a significant reduction in hypocotyl elongation (Supplemental Figure S10). We employed *rgi5x* to test whether the activity of CLEL6 as a negative regulator of anthocyanin accumulation depends on RGI receptors. In light-exposed *tpst-1* seedlings, anthocyanin concentration was reduced to wild-type levels when grown in presence of CLEL6, whereas anthocyanin levels in *rgi5x* were insensitive to CLEL6 treatment (Figure 5A). Also, under continuous light, hyperaccumulation of anthocyanins was reduced to wild- type levels by CLEL6 treatment in *tpst-1* but not in *rgi5x* (Figure 5B).

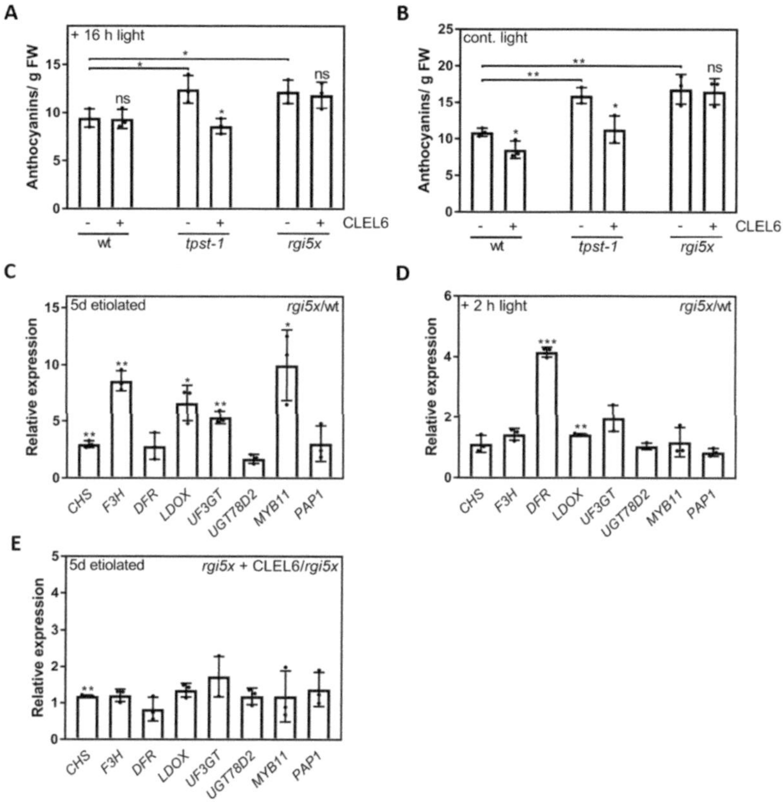

Figure 5: The *rgi5x* receptor mutant is insensitive to CLEL6 treatment. A Quantification of anthocyanin content of five-day-old etiolated wt, *tpst-1* and *rgi5x* seedlings after 16h light exposure. 1 µM CLEL6 was added to the media as indicated (-/+). **B** Quantification of anthocyanin content in five-day-old wt, *tpst-1* and *rgi5x* seedlings grown under continuous light. 1 µM CLEL6 was supplemented to the media as indicated (-/+).**C** Relative expression of *CHS*, *F3H*, *DFR*, *LDOX*, *UF3GT*, *UGT78D2*, *PAP1*, *MYB11* in 5-day-old etiolated *rgi5x* seedlings relative to the wt control. Gene expression was normalized to two reference genes. **D** Relative expression of *CHS*, *F3H*, *DFR*, *LDOX*, *UF3GT*, *UGT78D2*, *PAP1*, *MYB11* in five-day- old etiolated *rgi5x* seedlings compared to the wt control. Seedlings were exposed to light for 2 h before the sampling. Gene expression was normalized to two reference genes. **E** Relative expression of *CHS*, *F3H*, *DFR*, *LDOX*, *UF3GT*, *UGT78D2*, *PAP1*, *MYB11* in 5 day-old etiolated *rgi5x* seedlings grown on a media with 1 µM CLEL6 relative to the *rgi5x* control. Gene expression was normalized to two reference genes. Data are shown as the mean ±SD for n=3 (A-E) independent experiments. Asterisks indicate

significant differences (**P*<0.05, ***P*<0.01, ****P*<0.001; (A, E) two-tailed *t*- test; (B, C, D) one sample *t*-test).

Dependence of CLEL6 activity on RGI receptors was also observed at the level of gene expression. Similar to the *sbt6.1* and *tpst-1* mutants (Figure 2, C-F), the expression of anthocyanin biosynthesis genes was elevated in the *rgi5x* receptor mutant, both in dark-grown seedlings as well as after light exposure (Figure 5, C and D). The phenotype of etiolated *rgi5x* seedlings was again more similar to sbt6.1 than to *tpst-1*, in that induction levels of anthocyanin biosynthesis genes were lower than in *tpst-1*, and apparently not sufficient to result in significant pigment accumulation in darkness (Supplemental Figure S9B). In contrast to *tpst-1*, where we observed strong suppression of anthocyanin biosynthesis genes in response to CLEL6 (Figure 3D), CLEL6 treatment could not reduce the expression of these genes in the *rgi5x* receptor mutant (Figure 5E). Finally, the *rgi5x* receptor mutant showed insensitivity to CLEL6 treatment also with respect to hypocotyl elongation. The overexpression of CLEL6 was previously reported to cause an increase in hypocotyl length of etiolated seedlings (Ghorbani et al., 2016), which we confirmed in the *35S:CLEL6* transgenic line (Supplemental Figure S11A). Consistent with this finding, hypocotyl growth was also stimulated by CLEL6 treatment in etiolated wild-type and *tpst-1* seedlings, and the effect was much reduced in the *rgi5x* mutant (Supplemental Figure S11B). The data indicate that CLEL6 activity depends on RGIs, and they confirm that CLEL6-signaling through RGI receptor(s) controls anthocyanin formation by suppression of anthocyanin biosynthesis genes (Figure 6).

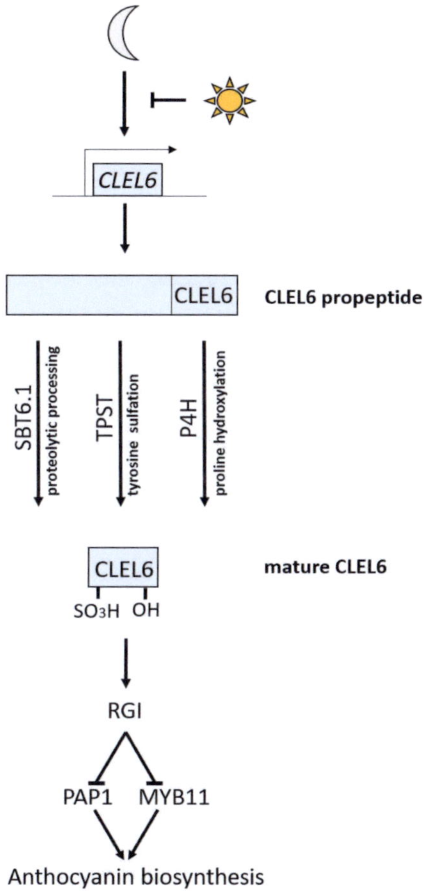

Figure 6: Model of CLEL6-regulated anthocyanin biosynthesis. During skotomorphogenesis, *CLEL6* is expressed to produce the CLEL6 precursor. Peptide biogenesis depends on SBT6.1 for proteolytic processing, on TPST for sulfation and on an unknown proline 4-hydroxylase (P4H) for proline hydroxylation. The mature CLEL6 peptide activates RGI receptor(s). Active RGI-signaling downregulates the expression of the PAP1 and MYB11 transcription factors. MYB11 induces the early biosynthesis genes while PAP1 induces the late biosynthesis genes for anthocyanin formation. The inhibitory effect of CLEL6 is attributed upon light perception, when *CLEL6* expression is downregulated to allow the accumulation of anthocyanins.

Discussion

CLEL6 is a small post-translationally sulfated peptide in the family of CLEL peptides, that are also known as ROOT GROWTH FACTORs (RGFs) or GOLVEN peptides (GLVs) (Matsuzaki et al., 2010; Meng et al., 2012; Whitford et al., 2012). These peptides are well-known for their function in root growth and development. CLELs control the size of the root meristematic zone and the stability of PLETHORA (PLT) transcription factors that are required for root stem cell maintenance (Yamada et al., 2020; Matsuzaki et al., 2010; Meng et al., 2012). CLELs are also involved in root gravitropism and have been implicated in lateral root development (Meng et al., 2012; Whitford et al., 2012). However, unlike most other CLELs that are predominantly expressed in roots, CLEL4 is induced in Arabidopsis shoots after pathogen infection (Wang et al., 2021), and CLEL6 and CLEL9 show constitutively high expression in hypocotyls and leaves (Whitford et al., 2012; Fernandez et al., 2013; Stegmann et al., 2022). Consistent with this expression pattern, CLEL6 and 9 are involved in growth and gravitropic responses of the hypocotyl (Whitford et al., 2012; Stührwohldt et al., 2020b; Ghorbani et al., 2016). The two peptides appear to act redundantly in these processes, and their formation and bioactivity depend on SBT6.1 for processing (Ghorbani et al., 2016; Stührwohldt et al., 2020b) and on TPST for post-translational sulfation (Stührwohldt et al., 2020b).

We report here an early photomorphogenic phenotype and mis-regulated anthocyanin biosynthesis for the *tpst* mutant, observed both in etiolated seedlings and in response to light, suggesting sulfated signaling peptides to be involved in the regulation of these processes. Mis-regulation of anthocyanin biosynthesis was also observed for the *sbt6.1* mutant that is impaired in the biogenesis of the CLEL peptides, particularly of CLEL6 and CLEL9 (Stührwohldt et al., 2020b). Confirming a role for CLEL peptides in the regulation of anthocyanin biosynthesis, the overaccumulation phenotype of *tpst* and *sbt6.1* mutants was suppressed by treatment with CLEL6. Furthermore, CLEL6 inhibited the expression of anthocyanin biosynthesis genes in etiolated as well as light-exposed seedlings, and its activity was lost in the *rgi5x* receptor mutant.

Interestingly, CLEL9 treatment or overexpression of the CLEL9 peptide precursor did not affect anthocyanin biosynthesis. Therefore, unlike hypocotyl growth and gravitropic responses that are similarly affected by both peptides, there is no redundancy between CLEL6 and CLEL9 with respect to the inhibition of anthocyanin biosynthesis. However, the observation that anthocyanin biosynthesis genes are upregulated to a much higher level in etiolated *tpst-1* compared to *sbt6.1* seedlings (Fig. 2B vs. Fig 2C) suggests that additional sulfopeptides may exist that suppress the expression of these genes and anthocyanin formation in the dark, and that do not rely on SBT6.1 for activity. Existance of additional negative regulators of anthocyanin biosynthesis is supported by the observation that loss of CLEL6 function by RNA interference in an artificial microRNA line (Whitford et al 2012) does not result in anthocyanin hyper-accumulation. The apparent differences in signaling specificity between CLEL6 and CLEL9 may be explained by differences in RGI-dependent signal perception. Our data show that CLEL6 is perceived redundantly by more than one as yet unidentified members of the RGI family, while RGI3 was characterized as the specific CLEL9 receptor (Stegmann et al., 2022). Comparing the amino acid sequence of CLEL6 (DYPQPHRKPPIHNE) and CLEL9 (DYNSANKKRPIHN), we noted three proline residues in CLEL6 that are missing from CLEL9, as a likely cause for differences in peptide structure, and thus, receptor binding properties. Differences in binding specificity may also result from the post-translational hydroxylation of prolines that are present only on CLEL6.

We propose CLEL6 as a negative regulator of anthocyanin biosynthesis that is active during skotomorphogenesis to suppresses anthocyanin formation in the dark. This inhibition is partially released when the expression of CLEL6 is reduced in the light (Figure 6, Supplemental Figure S3A and S3B). While anthocyanins accumulate in etiolated seedlings after light exposure, and in plants grown in continuous light, their level is still limited by CLEL6, suggesting that the peptide acts as a regulator of anthocyanin accumulation also in light grown plants (Supplemental figure S2, S7, S8A).

The onset of anthocyanin formation during seedling establishment is tightly controlled by transcriptional regulators of photomorphogenesis that inactivate (PIFs) or activate (HY5) anthocyanin biosynthesis genes (Shin et al., 2007; Lee et al., 2007). We hypothesize that they may do so by regulating CLEL6 expression. HY5 activates photomorphogenic genes including those for anthocyanin biosynthesis (Lee et al., 2007). HY5 also induces the expression of PAP1 which in turn stimulates late anthocyanin biosynthesis genes (Shin et al., 2013). Repression of CLEL6 expression in the light may constitute an additional mechanism by which HY5 could contribute to light-dependent anthocyanin formation.

Alternatively, expression of CLEL6 in etiolated seedlings and its downregulation after light exposure may be mediated by transcription factors like PIFs, that are inactivated by light. The PIF family comprises eight members and they are key regulators in the transition from skotomorphogenic development to photomorphogenesis. PIFs accumulate in darkness and are rapidly degraded in response to light (Duek and Fankhauser, 2005; Leivar et al., 2008). Most important for the shift from skoto- to photomorphogenesis are PIF1 and PIF3 (Shin et al., 2009; Shen et al., 2005). CLEL6 activity in etiolated seedlings matches that of PIFs. Like PIFs, CLEL6 promotes hypocotyl growth (Ghorbani et al., 2016); Supplemental Figure S11) and inhibits anthocyanin biosynthesis (Figures 3 and 4), and these activities are reduced when seedlings are exposed to light. CLEL6 may thus act downstream of PIFs, thereby contributing to the transition from skoto- to photomorphogenesis.

Evidence supporting the hypothesis that CLEL6 expression is controlled by PIFs was found in published sequence data. Reduced expression of CLEL6 in the *pifq* mutant compared to the wild type was reported in three independent RNA-seq studies (Martín et al., 2016; Ou et al., 2016; Richter et al., 2010). Two further studies reported lower expression of CLEL6 in a *pif4pif5* double mutant compared to wild type (Lorrain et al., 2008). In a promoter analysis by (Martín et al., 2016), CLEL6 was predicted as a direct target of PIF1, and (Oh et al., 2012) reported that PIF4 binds to the promoters of CLEL6 and TPST. Taken together, these data provide strong evidence for the hypothesis that the expression of CLEL6 in etiolated seedlings and its downregulation upon light exposure are controlled by the PIF transcription factors. The negative effect of CLEL6 on the expression of the MYB11 and PAP1 transcription factors and of the anthocyanin biosynthesis genes is thus released when PIFs are degraded, allowing for the synthesis and accumulation of photoprotective pigments in response to light.

METHODS

Plant materials and growth conditions

Plants were grown on 0.5 × Murashige and Skoog medium (Murashige and Skoog, 1962) supplemented with 1% sucrose and 0.38% gelrite. One µM CLEL6 (DY(SO3H)PQPHRKPPIHNE; Pepscan, Lelystad, The Netherlands), 1 µM unsulfated CLEL6 (DYPQPHRKPPIHNE; PepMic, Suzhou, China) or 1 µM CLEL9 (DMDY(SO3H)NSANKKRPIHN; SciLight Peptide, Beijing, China) were added as indicated. Arabidopsis seeds were sterilized for 15 minutes in 70% ethanol. Seeds were stratified for 2 days at 4°C. Germination was induced by a 4h-6h light treatment (120 µmol m−2 s−1 white light). Afterwards, plants were grown at 22 °C in darkness or under continuous light conditions. For light stress experiments, seedlings were grown for five days in the dark and transferred into constant light (120 µmol m−2 s−1 white light) for 2 or 16 hours as indicated.

Arabidopsis lines

All transgenic lines and mutants used for this study were described previously; *tpst-1* (Komori et al., 2009), *tpst-2* (Kang et al., 2014), *sbt6.1* (Rautengarten et al., 2005), *35S:CLEL6* and *35S:CLEL9* (Whitford et al., 2012), and *rgi5x* (Ou et al., 2016). The *rgi1* (SALK_020659), *rgi2* (SALK_039309C), *rgi3* (SALK_053167C), *rgi4* (SALK_052069C) and *rgi5* (SALK_014726C) seeds were received from NASC.

Quantification of hypocotyl length and cotyledon opening

Seedlings were grown for five-day in darkness. The cotyledons were scored as open when they were separated from each other. Hypocotyl length was determined using Image J.

Anthocyanin extraction and measurement

The protocol was adapted from (Nakata et al., 2013). Approximately 30 seedlings were harvested for each sample and fresh weight was determined. Samples were frozen in liquid nitrogen. Frozen samples were homogenized in a TissueLyser LT (Qiagen, Venlo, Netherlands). 200 µl of extraction buffer (45% methanol, 5% acetic acid) were added and samples were vortexed vigorously. The cell debris was removed by centrifugation at 13000 rpm in a microfuge at 4°C for 10 minutes. The supernatant was transferred into a new microfuge tube. Centrifugation and transfer into a new tube were repeated. The Absorption was measured at 530 nm and 657 nm in a TECAN Spark plate reader (Tecan Deutschland GmbH, Crailsheim, Germany). Anthocyanin content was calculated as ([Abs530 - (0.25 x Abs657)] x 2) / g FW.

qPCR

Total RNA was extracted using TRIzol (ABP Biosciences, Beltsville, Maryland, USA) following the manufacturers protocol. Up to 5 µg RNA was used for cDNA synthesis by RevertAid reverse transcriptase (Thermo Fisher Scientific, Waltham, Massachusetts, USA) using oligo dT Primers. qPCR Primers are listed in Supplemental Table S1. qPCR was performed in a CFX96 Real-Time PCR Detection system (Bio-Rad, Hercules, California, USA). Experiments were performed in three biological and two technical replicates. Expression data were normalized against the two housekeeping genes *ACTIN2* (At3g18780) and *TUB4* (At5g44340).

Acknowledgements

Authors thank Lhana Stein for excellent technical support. We also thank Prof. Liu for the *tpst-2* line, Ana Fernandez for the *35S:CLEL6* and *35S:CLEL9* line and Prof. Li for the *rgi5x* line.

Supplemental figures

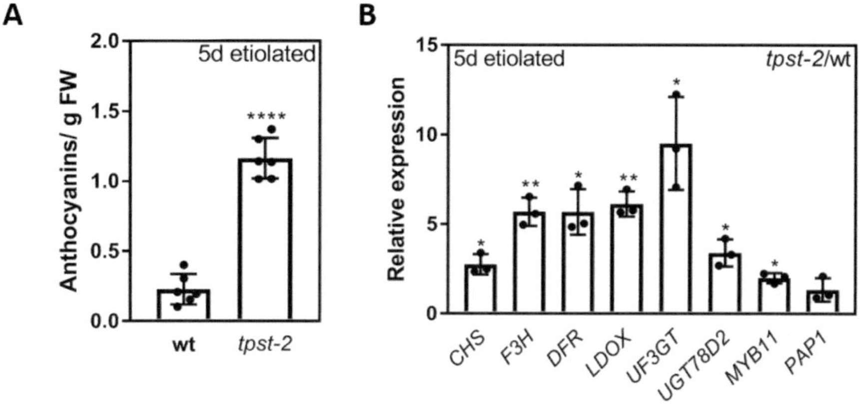

Supplemental Figure S1. Anthocyanin accumulation and biosynthesis genes are upregulated in the independent *tpst-2* mutant. A Quantification of anthocyanin content in five-day-old etiolated wild-type (wt) and *tpst-2* seedlings. **B** Relative Expression of *CHS*, *F3H*, *DFR*, *LDOX*, *UF3GT*, *UGT78D2*, *PAP1*, *MYB11* in five-day-old etiolated *tpst-2* seedlings relative to the wt control. Gene expression was normalized to two reference genes. Data are shown as the mean ±SD for n=6 (A) and n=3 (B) independent experiments. Asterisks indicate significant differences (**P*<0.05, ***P*<0.01, *****P*<0.0001; (A) two-tailed *t*-test; (B) one sample *t*-test).

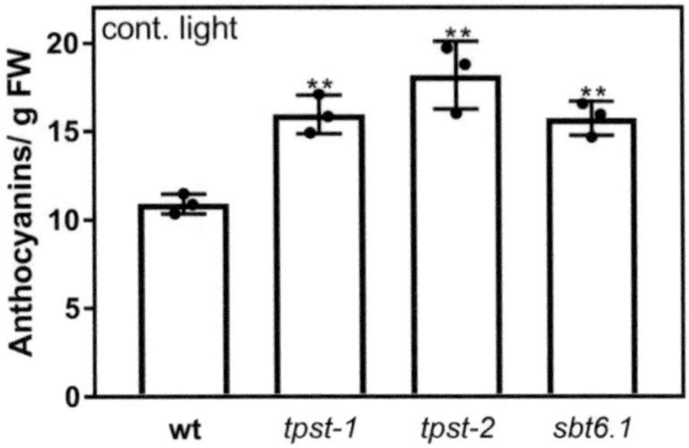

Supplemental Figure S2. *tpst*-mutants and *sbt6.1* accumulate significantly more anthocyanin in continuous light. Quantification of anthocyanin content in five-day-old wt, *tpst-1*, *tpst-2* and *sbt6.1* seedlings grown under continuous light. Data are shown as the mean ±SD for n=3 independent experiments. Asterisks indicate significant differences (**P<0.01; two-tailed t-test).

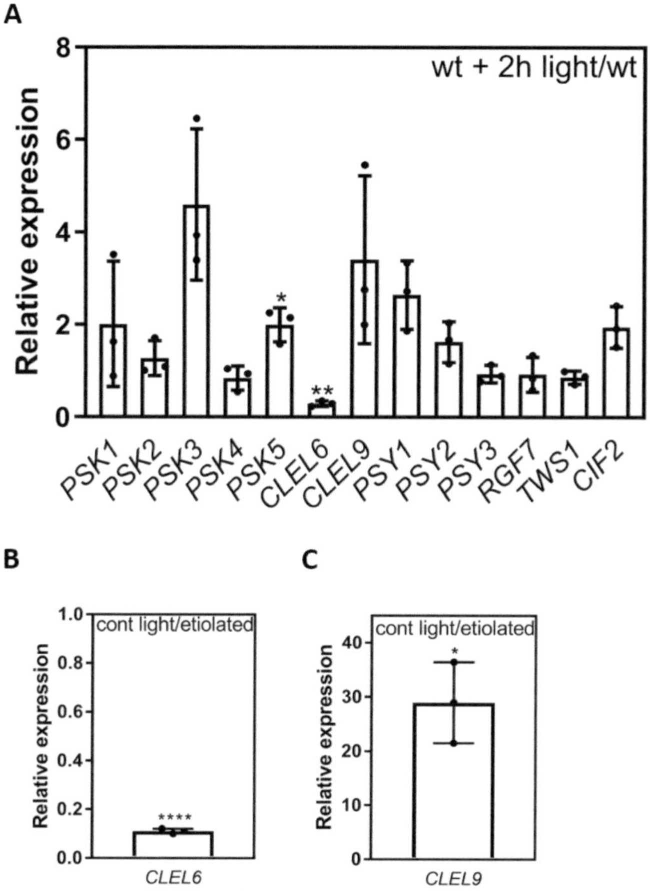

Supplemental Figure S3. *CLEL6* expression is downregulated by light. A Expression of the genes coding for sulfated peptides in five-day-old etiolated wild-type wt seedlings after 2h of light exposure relative to the dark control. Gene expression was normalized to two reference genes. Genes for which transcripts were not detectable were omitted from the graph. **B, C** Relative Expression of *CLEL6* and *CLEL9* in five-day-old wt seedlings grown under continuous light relative to the etiolated wt control. Gene expression was normalized to two reference genes. Data are shown as the mean ±SD for (A, B, C) n=3 independent experiments. Asterisks indicate significant differences (**P*<0.05, ***P*<0.01, *****P*<0.0001; one sample *t*-test).

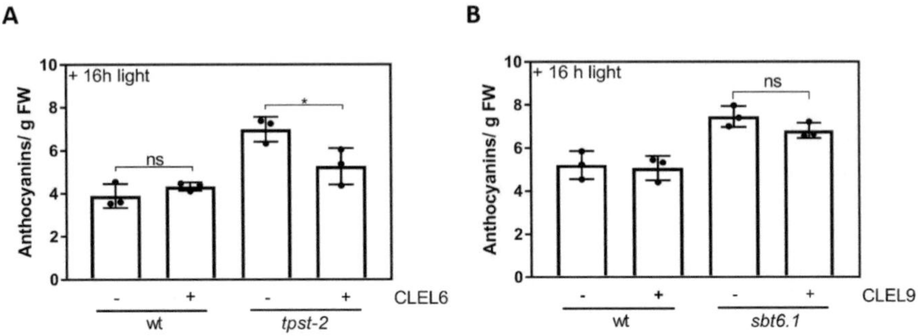

Supplemental Figure S4. CLEL6 complements the anthocyanin accumulation phenotype, CLEL9 does not. A Quantification of anthocyanin content in five-day-old etiolated wild-type (wt) and *tpst-2* seedlings after 16 h of light exposure. 1 µM CLEL6 was added as indicated (-/+). **B** Quantification of anthocyanin content in five-day-old wt and *sbt6.1* seedlings 16 h after exposure to light 1 µM CLEL9 was added as indicated (-/+). (A, B) Data are shown as the mean ±SD for n=3 independent experiments. Asterisks indicate significant differences (*P<0.05; two-tailed t-test).

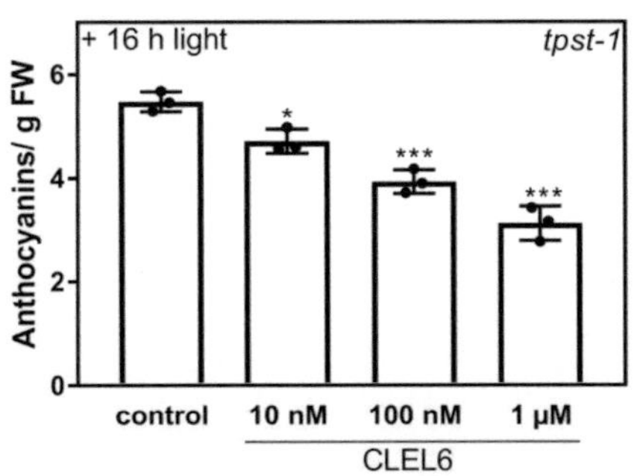

Supplemental Figure S5. Concentration-dependent reduction of anthocyanin content in *tpst-1* by CLEL6. Quantification of anthocyanin content in five-day-old etiolated *tpst-1* seedlings prior and after 16 h light exposure. 10 nM, 100 nM or 1 µM CLEL6 was added to the media as indicated. Data are shown as the mean ±SD for n=3 independent experiments. Asterisks indicate significant differences (*P<0.05, ***P<0.001; two-tailed t-test).

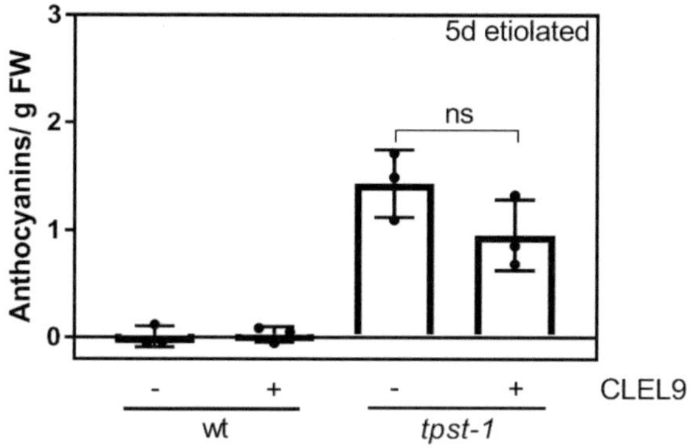

Supplemental Figure S6. CLEL9 has no effect on the anthocyanin content in *tpst-1*. Quantification of anthocyanin content in five-day etiolated wild-type (wt) and *tpst-1* seedlings with (+) or without (-) 1 µM CLEL9 in the media. Data are shown as the mean ±SD for n=3 independent experiments.

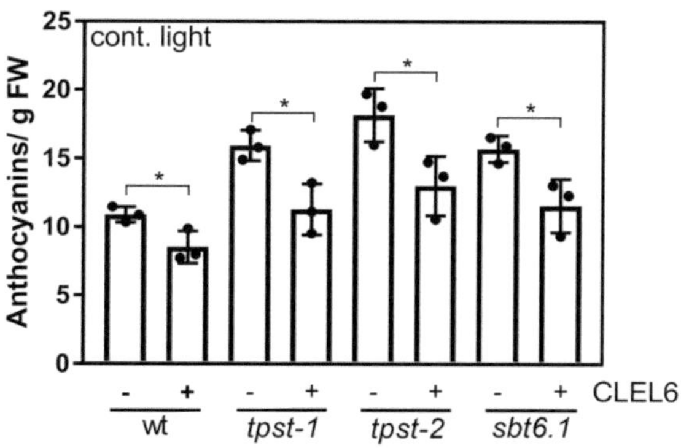

Supplemental Figure S7. CLEL6 inhibits anthocyanin biosynthesis in continuous light in wild-type, *tpst* and *sbt6.1* seedlings. Quantification of anthocyanin content in five-day-old wt, *tpst-1*, *tpst-2* and *sbt6.1* seedlings grown under continuous light. 1 µM CLEL6 was added to the media as indicated. Data are shown as the mean ±SD for n=3 independent experiments (* $P<0.05$; two-tailed *t*-test).

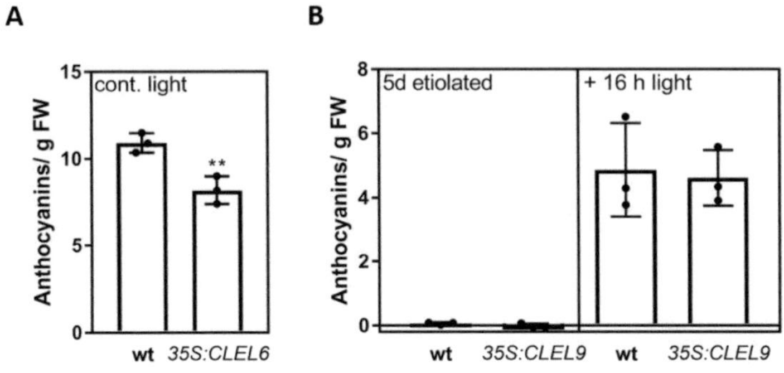

Supplemental Figure S8. Anthocyanin content in *35S:CLEL6* and *35S:CLEL9* plants. **A** Quantification of anthocyanin content in five-day-old wild-type (wt) and *35S:CLEL6* seedlings grown under continuous light. **B** Quantification of anthocyanin content in five-day-old etiolated wt and *35S:CLEL9* seedlings prior and after 16 h light exposure. Data are shown as the mean ±SD for n=3 independent experiments. Asterisks indicate significant differences (***P*<0.01; two-tailed *t*-test).

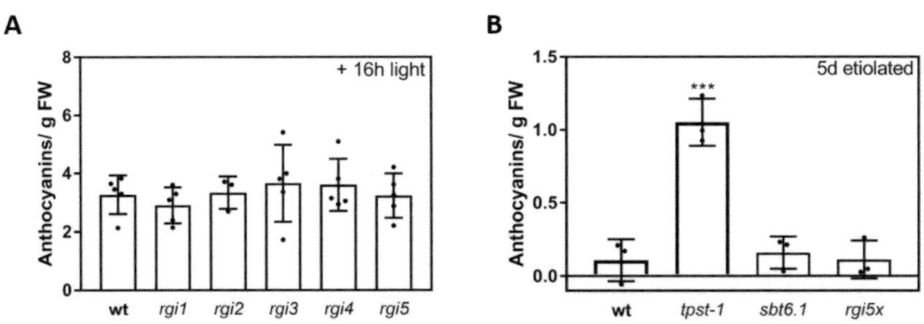

Supplemental Figure S9. Etiolated *rgi5x* seedlings do not accumulate anthocyanins. **A** Quantification of anthocyanin content in five-day-old etiolated wild-type (wt), *rgi1*, *rgi2*, *rgi3*, *rgi4* and *rgi5* seedlings after 16 h of light exposure. **B** Quantification of anthocyanin content in five-day-old etiolated wt, *tpst-1*, *sbt6.1* and *rgi5x* seedlings. Data are shown as the mean ±SD for (A) n=5 (*rgi2*, n=3) and (B) n=3 independent experiments. Asterisks indicate significant differences (****P*<0.001; two-tailed *t*-test).

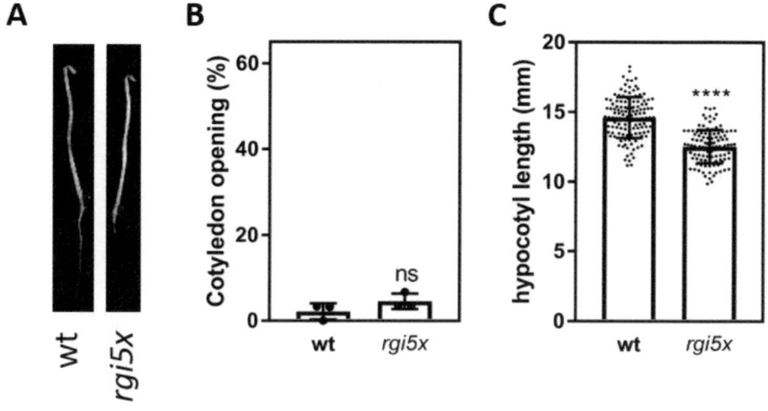

Supplemental Figure S10. Phenotype of dark-grown *rgi5x* seedlings. A Representative five-day-old etiolated wild-type (wt) and *rgi5x* seedlings. **B, C** Quantification of **(B)** open cotyledons (in %) and **(C)** hypocotyl length (in mm) of five-day-old etiolated wt and *rgi5x* seedlings. Data are shown as the mean ± SD. (B) n=3 independent experiment with at least 27 seedlings each. (C) n=3 independent experiment with at least 26 seedlings each. Asterisks indicate significant differences (*****P*<0.0001; two-tailed *t*-test).

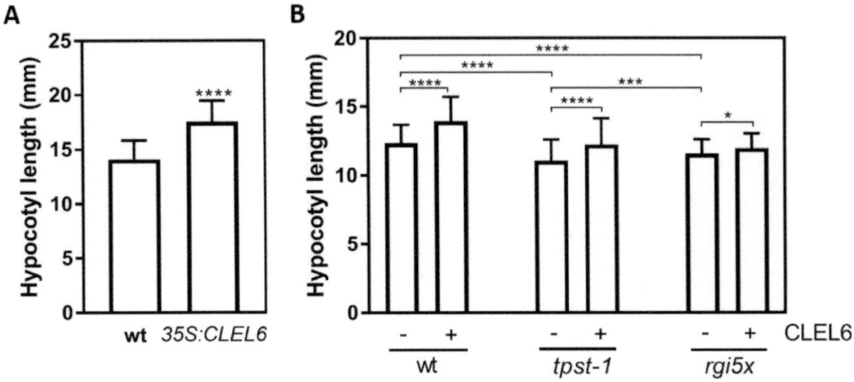

Supplemental Figure S11. CLEL6-induced hypocotyl growth depends on RGI receptors. A Quantification of hypocotyl length of five-day-old etiolated wild-type (wt) and *35S:CLEL6* seedlings. **B** Quantification of hypocotyl length of five-day-old etiolated wt, *tpst-1* and *rgi5x* seedlings grown with (+) or without (-) 1 µM CLEL6 in the media. Asterisks indicate significant differences between the mutants and wt, and between CLEL6-treated and untreated plants. (A, B) Data are shown as the mean ±SD for n=3 independent experiments. Asterisks indicate significant differences (*P<0.05, ****P<0.0001; two-tailed t-test).

Supplemental Table S1. qPCR Primer.

Accession No.	Name	Sequence (5'-3')	Publication
At3g18780	*ACTIN2* RT F	GTGGATATCAGGAAGGATCTGTACGGT	
At3g18780	*ACTIN2* RT R	TTCTGTGAACGATTCCTGGACCTGC	
At5g44340	*Tub4* RT F	TGAGGAAGGAAGCTGAGAACAG	
At5g44340	*Tub4* RT F	ATACACTGGTCAGCGTTTTC	
At5g13930	*CHS* RT F	CCTGACTACTACTTCCGCATCACC	
At5g13930	*CHS* RT R	CGCACATGCGCTTGAACTTCTC	
AT3G51240	*F3'H* RT F	CGGTGGACTGGGCTATAGCTGA	Zhao et al., 2021
AT3G51240	*F3'H* RT R	CGAGAGTGGTGTTGGTGGATG	Zhao et al., 2021
AT5G42800	*DFR* RT F	CTTTGTTCGTGCCACCGTTCG	Zhao et al., 2021
AT5G42800	*DFR* RT R	AAAATCCATGGGTGTTGCCAC	Zhao et al., 2021
AT4G22880	*LDOX* RT F	GTTTGCAGCTTTTCTACGAGGGC	Zhao et al., 2021
AT4G22880	*LDOX* RT R	ATGTTGAGCAAAAGTCCGTGGAG	Zhao et al., 2021
AT5G54060	*UF3GT* RT F	TTGTCAGATCGTTTTGGTTCCGC	Zhao et al., 2021
AT5G54060	*UF3GT* RT R	TCTTCCTCACTTTCTCACCGATC	Zhao et al., 2021
AT5G17050	*UGT78D2* RT F	CGGTGTTGGAGAGTGTATCGG	Zhao et al., 2021
AT5G17050	*UGT78D2* RT R	CCAATCTCCCACACAACCTCC	Zhao et al., 2021
AT1G56650	*PAP1* RT F	TGGTTCCTGAAGCGACGACAAC	Zhao et al., 2021
AT1G56650	*PAP1* RT R	CGCAAACAAATGTTCGAAACAC	Zhao et al., 2021
AT3G62610	*MYB11* RT F	GATGGCGATTGTAACCCAAGC	Zhao et al., 2021
AT3G62610	*MYB11* RT R	ACATGAGGACACGTGGACAGC	Zhao et al., 2021
AT4G16515	*CLEL6* RT F	TCCAATGTTGGATGTGCCTCTGC	
AT4G16515	*CLEL6* RT R	TCGTTGTGGATCGGAGGCTTAC	
AT5G64770	*CLEL9* RT F	ACAAGTCTTTCCTCGTCGCACTTC	
AT5G64770	*CLEL9* RT R	ACTTCTCGTAAGCTTCGTGCTTG	
AT3G02240	*RGF7* RT F	ACGGGCAACGAAAGTACAAACCC	
AT3G02240	*RGF7* RT R	TGACTACGTACTACGCGCAACC	
AT5G58650	*PSY1* RT F	CAACCCTGTTTCCGTTTCAGGTG	
AT5G58650	*PSY1* RT R	TTGCAGATGGGTCACCGTAGTC	
AT3G47295	*PSY2* RT F	TGACGATCCATCTGCAAACACCAG	
AT3G47295	*PSY2* RT R	AAGGTGTTGTGTCAGCCTTTGC	
AT2G29995	*PSY3* RT F	TCTCTCTTCTGCTCGCATCAGTC	
AT2G29995	*PSY3* RT R	TCAATGACCGTTCTGGCACCAC	
AT5G01075	*TWS1* RT F	AGCAATCCCAAGACGGTGGATAAC	
AT5G01075	*TWS1* RT R	ACAGTGCGTAGATCACATTCACG	
AT4G34600	*CIF2* RT F	GATGGGTTTGTTGCCATTGGTG	
AT4G34600	*CIF2* RT R	CAGAGAGCAAACGCTGAAGCTG	
AT1G13590	*PSK1* RT F	ACAGTAGCTGCTCACACCGATTAC	Stührwohldt et al., 2021
AT1G13590	*PSK1* RT R	AGAGCTGAGTTTGAGGGCGAGAAG	Stührwohldt et al., 2021
AT2G22860	*PSK2* RT F	TGCTCCACGCTAATGTGCACTG	Stührwohldt et al., 2021
AT2G22860	*PSK2* RT R	ATCTTCGTCGTCTGCACCACAG	Stührwohldt et al., 2021
AT3G44735	*PSK3* RT F	AAGCTTGTGCCTGGCAGTTCTC	Stührwohldt et al., 2021
AT3G44735	*PSK3* RT R	TGAGTCCTCTTCCACTGATGTAGC	Stührwohldt et al., 2021
AT3G49780	*PSK4* RT F	TACGCTAACCTACGCAGCAAGG	Stührwohldt et al., 2021
AT3G49780	*PSK4* RT R	TCTCCAATTCCGTTGCAGCTTTC	Stührwohldt et al., 2021
At5g65870	*PSK5* RT F	CTCTGCTCCACGCTAACACA	Stührwohldt et al., 2021
At5g65870	*PSK5* RT R	TCTTCTCCAACACCTTCACAGA	Stührwohldt et al., 2021

2.2 Tyrosylprotein Sulfotransferase activity is required for photomorphogenesis and starch production

Running title: TPST-regulated photomorphogenesis

Eric Bühler[1], Lim Magdiel[2], Urska Repnik[3], Till Ischebeck[2], Andreas Schaller[1] and Nils Stührwohldt[1]

[1] Department Plant Physiology and Biochemistry, Institute of Biology, University of Hohenheim, 70593 Stuttgart, Germany.

[2] Institute of Plant Biology and Biotechnology (IBBP), Green Biotechnology, University of Münster, 48143 Münster, Germany

[3] Central Microscopy, Department of Biology, University of Kiel, 24118 Kiel, Germany

Status: unpublished manuscript

Own contribution: I contributed to the study design and performed all experiments except the transmission electron microscopy (Figure 4A) and GC-MS measurements (Figure 4D, Supplemental figure S5, Supplemental table S1). I wrote the manuscript with help from Nils Stührwohldt and Andreas Schaller.

Abstract

Photomorphogenesis is a crucial step in plant development and is regulated by many factors. Here we identify a role for the Tyrosylprotein Sulfotransferase (TPST) enzyme in regulating the photomorphogenesis and starch production. TPST catalyzes the essential tyrosine sulfation of post-translationally modified peptides. Etiolated *tpst*-mutant seedlings showed an early photomorphogenesis phenotype characterized by a shorter hypocotyl, open cotyledons, high accumulation of protochlorophyllide, higher transcript amounts of several photomorphogenesis-related genes and disturbed etioplast formation. By assaying protochlorophyllide content, qPCR analysis and feeding assays with 5-aminolevulinic acid (ALA), we demonstrate increased protochlorophyllide synthesis in *tpst*-mutants. We further show early photomorphogenesis of the *tpst*-mutant may be caused by a higher expression level of the photomorphogenesis inducing transcription factor GOLDEN-2-LIKE (GLK1). A similar behavior was observed for *35S:GLK1* overexpressors and the *tpst*-mutant, since both are able to de-etiolate when retrograde signaling was activated by lincomycin treatment. Furthermore, a TPST dependent signal regulates starch accumulation, as indicated by electron transmission microscopy that showed a large starch accumulation in *tpst*-mutant etioplasts. Taken together our data indicate that TPST has an important role in retaining the skotomorphogenic development by inhibiting the expression of GLK1. Additionally, we propose a role for TPST in regulating the starch production that has to be studied further.

Introduction

Photomorphogenesis represents the largest developmental change in a plant life cycle. Etiolated seedlings are characterized by a long hypocotyl, closed cotyledons and an apical hook. After exposure to light the cotyledons open and chloroplasts and the photosynthetic apparatus are formed. Many external and internal factors are involved in this developmental transition.

In etiolated seedlings two major regulators retain the skotomorphogenic growth, the PHYTOCHROME INTERACTING FACTOR (PIF) transcription factors and the CONSTITUTIVE PHOTOMORPHOGENIC- SUPPRESSOR OF PHYA-105 (COP1-SPA) E3 ligase complex. The COP1-SPA complex inhibits photomorphogenesis by degradation of transcription factors like ELONGATED HYPOCOTYL 5 (HY5) and LONG HYPOCOTYL IN FAR-RED1 (HFR1) (Osterlund et al., 2000; Jang et al., 2005). The PIFs are a group of eight basic helix-loop-helix transcriptional regulators that control photomorphogenesis by inhibiting the expression of a broad range of photomorphogenic genes, including genes coding for enzymes catalysing the chlorophyll synthesis (for example *HEMA*, *GUN4*) or for light harvesting complexes (*LHC*) (Duek and Fankhauser, 2005). Light-activated phytochrome leads to a reorganization of the COP1-SPA complex and to degradation of the PIFs, resulting in photomophogenic development (Leivar et al., 2008; Jiao et al., 2007; Sheerin et al., 2015).

To do photosynthesis, plants need to form chloroplasts and chlorophyll. In etiolated seedlings the biogenesis of chloroplasts is halted at the stage of etioplasts and chlorophyll synthesis stops at the intermediate protochlorophyllide (pchlide) (Mullet, 1988; Sakamoto et al., 2008). The conversion from pchlide to chlorophyllide, which is catalyzed by the enzyme protochlorophyllide oxidoreductase (POR), requires light energy (Armstrong et al., 1995).

An important transcription factor class regulating chloroplast biogenesis are the GOLDEN-2-LIKEs (GLKs). They integrate retrograde and anterograde signaling (Martín et al., 2016). Anterograde signaling describes a signal that is sent from the nucleus to the chloroplasts, while retrograde signaling describes the reversed pathway of signals. In the case of a disturbed chloroplast the retrograde signaling is activated, the GLKs and thereby the expression level of several photomorphogenic genes are downregulated through GENOMES UNCOUPLED1 (GUN1) to avoid further damage (Koussevitzky et al., 2007).

Previous studies showed that a mutation in the *TPST* gene, which encodes for the Tyrosylprotein Sulfotransferase, leads to an early photomorphogenesis phenotype in etiolated seedlings, including a shorter hypocotyl and open cotyledons (Kang et al., 2014, Giorgi et al., 2021). TPST is catalyzing tyrosine sulfation which is a posttranslational modification on proteins and peptides that occurs all over eukaryotes by utilizing the cosubstrate 3′-phosphoadenosine 5′-phosphosulfate (PAPs) as sulfate donor (Moore, 2003). However, plant and animal TPSTs show no sequence similarity, suggesting that they evolved through convergent evolution and may differ in function (Komori et al., 2009). In Arabidopsis, TPST was characterized as a cis-Golgi localized transmembrane protein (Komori et al., 2009). Loss of TPST function leads to dwarfed plants with pale green leaves and stunted roots (Komori et al., 2009).

In this study we show that TPST activity is required to retain skotomorphogenesis and to promote a correct transition into photomorphogenic growth. TPST enhances hypocotyl elongation, and inhibits cotyledon opening and pchlide synthesis in etiolated seedlings. A loss of TPST function leads to high pchlide accumulation resulting in photobleaching after transfer to light. We also show that TPST is required to regulate the accumulation of starch in etioplasts through sugar transport.

Results

Etiolated *tpst*-mutant seedlings accumulate more pchlide, which causes photobleaching upon transfer to light

Since PIFs are negative regulators of photomorphogenesis, the *pifq* quadruple mutant lacking PIF1, PIF3, PIF4 and PIF5 has a constitutive photomorphogenic (*cop*)-like phenotype in the dark (Leivar et al., 2008). Etiolated *tpst*-seedlings display a shorter hypocotyl and open cotyledons, which was described earlier by Giorgi et al., 2021 and Bühler et al., 2023 and verified further in our experiments by using two individual *tpst*-mutant lines (Fig. 1A, Suppl. Fig. S1A), thus phenocopying the *pifq* mutant. The *cop*-like phenotype of *pifq* is suppressed when retrograde signaling is activated by lincomycin (linc), a drug that inhibits protein biosynthesis in plastids and thus interferes with chloroplast development (Martín et al., 2016). Therefore, to confirm the phenotype of *tpst-1* as *cop*-like, we treated etiolated seedlings with linc and observed that early opening of cotyledons was indeed suppressed (Fig. 1A). Indicating that the TPST is needed for correct photomorphogenesis.

Next, we further analysed the *tpst* early photomorphogenesis phenotype using two independent loss-of-function alleles, *tpst-1* and *tpst-2* (*tpst-2* always shown in the supplemental data). After etiolated seedlings are exposed to light, the greening sets in, concomitant with chloroplast formation and chlorophyll synthesis. Four to six days old etiolated *tpst*-mutant seedlings show reduced greening rates after the exposure to light (Fig. 1B, C, Suppl. Fig. S1B). Greening rate of the *tpst*-mutant decreases with each additional day in the dark and the cotyledons display white photobleached regions (Fig. 1B, C). Previously, it was shown that photobleaching events during greening may be caused by the photooxidative damage triggered by the precursor of chlorophyll, pchlide, which accumulates in the dark (Meskauskiene et al., 2001). Five-day etiolated *tpst*-mutant seedlings accumulated significantly more pchlide than the WT (Fig. 1D, Suppl. Fig. S1C), which indicates a high potential for photooxidative damage upon light exposure. In the light excess pchlide is converted into chlorophyll, and a lack of CHLOROPHYLL A/B BINDING PROTEINS (CABs) to bind the free chlorophyll will cause the photooxidative damage (Meskauskiene et al., 2001). The EARLY LIGHT INDUCEBLE PROTEIN (ELIPs) are expressed very early during greening to help binding the free chlorophyll and reduce the photooxidative damage (Hutin et al., 2003). The relative expression levels of *ELIP1* and *ELIP2* were significantly increased in the *tpst*-mutants (Suppl. Fig. S1D). Thereby the *tpst*-mutants have a higher capacity of binding free chlorophylls during greening, which, however, is still not sufficient to prevent bleaching. A further qPCR analysis showed the increase in pchlide correlated with higher expression of genes involved in pchlide synthesis in *tpst*-mutants (Fig. 1E, Suppl. Fig. S1E). Except *HEMA*, *GUN4* and *CAO* all tested genes were significantly upregulated in etiolated *tpst*-mutant seedlings in comparison to the WT (Fig. 1E, Suppl. Fig. S1E). In contrast to the pchlide synthesis related genes, the expression level of genes involved in the partially shared phytochromibilin synthesis pathway *FC1*, *HY1* and *HY2*, were not differentially regulated in the *tpst-1* mutant (Suppl. Fig. S2A).

Because pchlide is a photosensitizer and can cause photooxidative damage upon exposure to light, the accumulation has to be regulated by many factors. The FLU protein inhibits glutamyl-tRNA reductase which catalyses the first step of pchlide synthesis, resulting in 5-aminolevulinic acid (ALA) (González-Carranza et al., 2012). In the *flu* mutant this inhibitory block is missing, resulting in the accumulation of large amounts of pchlide in etiolated seedlings (Meskauskiene et al., 2001). In feeding experiments with ALA, we observed that the *tpst-1* mutant produces 7.2 times more protochlorophyllide, while the WT only produces 4.3 times more protochlorophyllide out of the same amount of ALA (Fig. 1F). This result is consistent with the upregulation of the entire pathway observed at the level of gene expression (Fig. 1E, Suppl. Fig. S1E). Since the *tpst*-mutant shows a broad defect in photomorphogenesis, we hypothesize an altered activity of a general regulator of photomorphogenesis in *tpst*-mutants.

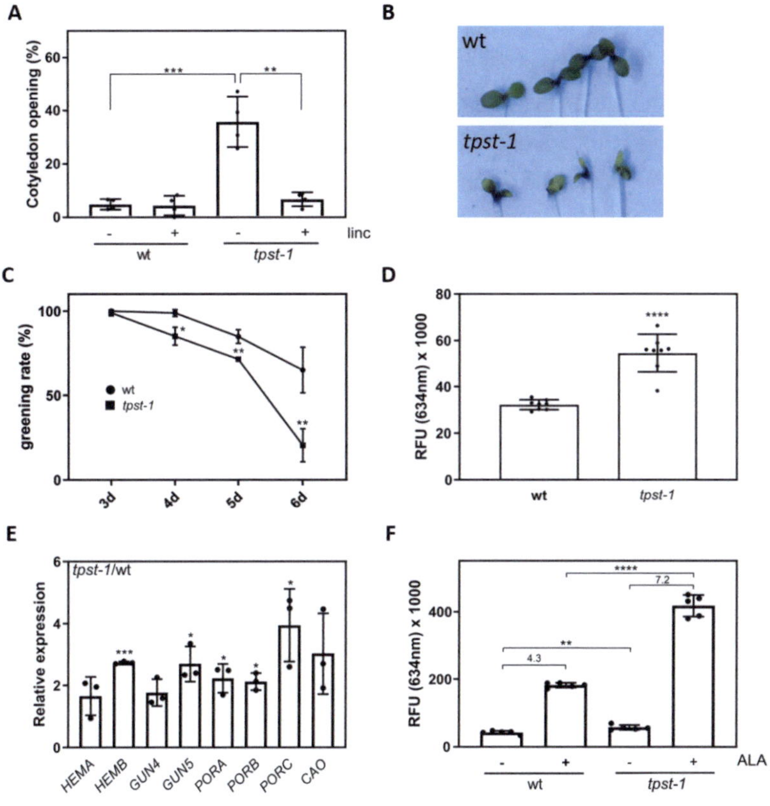

Figure 1. High accumulation of pchlide in etiolated *tpst-1* seedlings causes reduced greening rates. **A** Quantification of the open cotyledons of five-day etiolated wt and *tpst-1* seedlings treated with or without 0.5 mM lincomycin (linc) in the media. **B** Representative five-day etiolated WT and *tpst-1* seedlings after transfer to two days of constant light. **C** Determination of green cotyledons. Percentage of open cotyledons is shown for three to six-day etiolated seedlings observed after transfer to two days of constant light. **D** Pchlide measurement of five-day etiolated seedlings. The concentration of pchlide is shown in relative fluorescence units (RFU) at 634 nm. **E** Relative expression level of *HEMA*, *HEMB*, *GUN4*, *GUN5*, *PORA*, *PORB*, *PORC* and *CAO* in five-day etiolated seedlings compared to the WT control. Gene Expression was normalized against two reference genes (*ACTIN2* and *TUB4*). **F** Pchlide measurement of five-day etiolated seedlings with or without 100 µM ALA in the media. Data are shown as the mean ±SD: (A) n=4, (C) n=3, (D) n=8 (E) n=3, (F) n=5. Asterisks indicate significant differences (*P<0.05, **P<0.01, ****P<0.0001; (A, C, D, F) two-tailed t-test; (E) one sample t-test).

The *tpst-1* mutant has a higher PIF1 protein abundance

The dark accumulating PIFs are a well-known family of bHLH transcription factors regulating photomorphogenesis. PIF1 and PIF3 are the most dominant PIFs for inhibiting the photomorphogenesis in darkness (Shen et al., 2005a; Leivar et al., 2008). Since *tpst*-mutants phenocopy the *pifq* mutant we analyzed the PIF protein and expression levels. *PIF1*, *PIF3* and *PIF4* expression levels in the *tpst-1* mutant were unchanged compared to wildtype levels (Suppl. Fig. S3A). The PIFs are not only regulated on the gene expression level, but also on the protein level through many proteins like phytochromes, COP1/SPA and DELLAs (Shen et al 2005, Shin et al 2009, Pham et al 2018, Bauer et al 2004, Li et al 2016). The amount of PIF3 in the *tpst-1* mutant is similar to WT, while PIF1 protein abundance is even higher in etiolated *tpst-1* than in WT seedlings (Fig. 2A, B, C). Therefore, the early photomorphogenesis phenotype of *tpst* is not caused by the lack of the negative regulators of photomorphogenesis, PIF1 and PIF3. We rather hypothesize a positive regulator that overcomes the negative effect of PIF1 and PIF3.

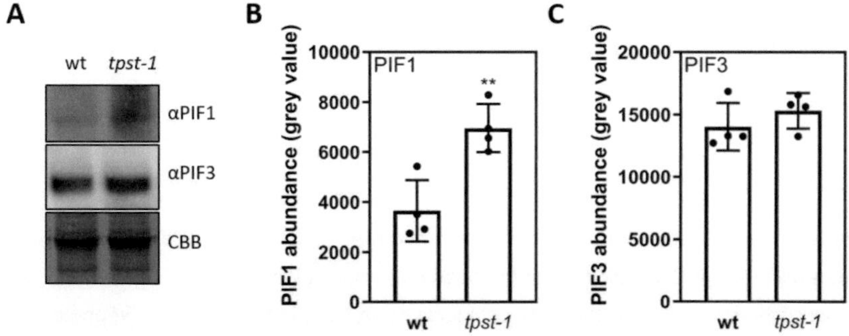

Figure 2. *tpst-1* has a higher abundance of PIF1. A Western blot analysis of PIF1 and PIF3 protein level in five-day etiolated WT and *tpst-1* seedlings. Coomassie brilliant blue staining as a loading control. **B** Quantification of the PIF1 protein level. **C** Quantification of the PIF3 protein level. Band intensis is shown as grey value. Data are shown as the mean ±SD: (A, B, C) n=4. Asterisks indicate significant differences (**P<0.01; two-tailed t-test).

The *tpst-1* mutant acts like a GLK1 overexpressing line

One important class of transcription factors positively regulating photomorphogenesis are the GLKs. The GLK family consists of two members and they play a crucial role in chloroplast biogenesis (Waters et al., 2009; Kobayashi et al., 2013). The expression level of *GLK1* is significantly higher in the etiolated *tpst*-mutant, while *GLK2* expression was not affected (Fig. 3A, Suppl. Fig. S4A). Suggesting that the early photomorphogenesis phenotype of the *tpst*-mutants is caused by a high expression level of GKL1. A previous study showed that a GLK1 overexpressing line is able to de-etiolate under weak light (1 µmol m^{-2} s^{-1}), when retrograde signaling is activated by lincomycin in the media (Martín et al., 2016). Likewise, the *tpst*-mutant showed partial de-etiolation in presence of lincomycin (Fig. 3B, C), illustrating that the *tpst*-mutant acts like a GLK1 overexpressing line. Consistent with this notion, the expression level of the GLK1 downstream targets *LHCA1*, *LHCA2*, *LHCB1.1* and *LHCB2.1* was significantly upregulated in etiolated *tpst-1* seedlings (Fig 3D).

Since the overexpression of LHCB1 and LHCB2.1 is known to promote etioplast differentiation (Liu et al., 2017), and prolamellar body formation is impaired in the *glk1/glk2* double mutant (Waters et al., 2009), we next investigated etioplast formation in *tpst*.

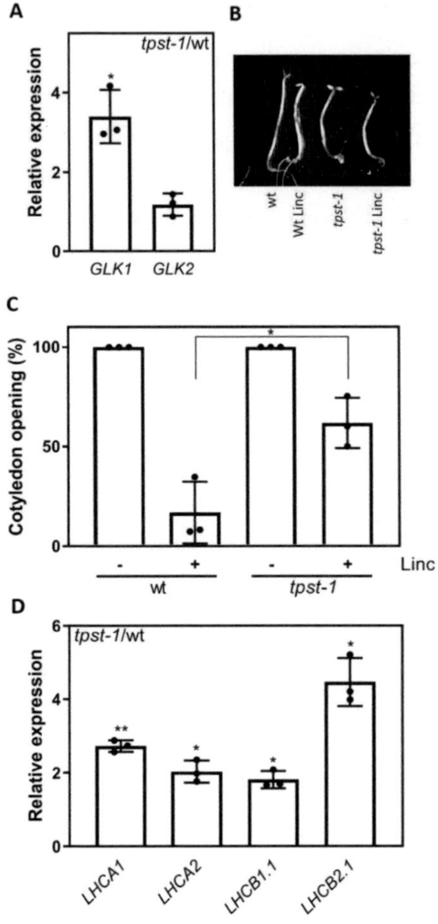

Figure 3. *tpst-1* acts like an GLK1 overexpressing line. A Relative expression level of *GLK1* and *GLK2* in five-day etiolated seedlings compared to the WT control. Gene Expression was normalized against two reference genes. **B** Representative seedlings grown under weak light (1 µmol m^{-2} s^{-1}) for four days. **C** Quantification of the open cotyledons of WT and *tpst-1* seedlings grown three days under weak light with or without 0,5 mM lincomycin in the media. **D** Relative expression level of *LHCA1*, *LHCA2*, *LHCB1.1*, *LHCB2.1* in five-day etiolated seedlings compared to the WT control. Gene Expression was normalized against two reference genes. Data are shown as the mean ±SD: (A) n=3, (C) n=3, (D) n=3. Asterisks indicate significant differences (*P<0.05, **P<0.01, ***P<0.001, ****P<0.0001; (C) two-tailed t-test; (A, D) one sample t-test).

tpst-1 etioplasts are disrupted by a high accumulation of starch

Transmission electron microscopy (TEM) showed that etioplast ultrastructure is strongly affected in *tpst-1* mutants. In WT etioplasts the prolamellar body is well established in symmetrically arranged tubular tetrahedral units. The most striking difference in the *tpst-1* etioplasts is the large number of starch grains, that in some cases disturbes prolamellar body formation (Fig. 4A, Fig. 4B). Moreover, notable differences between WT and *tpst-1* TEM includes *tpst-1* mutant lacking visible vacuoles and displaying black dots that might be associated with lipid accumulation. Consistent with the overaccumulation of starch, qPCR analysis of etiolated *tpst-*

mutants showed a higher expression level of the *APL3* gene, which encodes for a subunit of the ADP-Glucose pyrophosphorylase catalyzing the first and rate limiting step of starch synthesis (Fig. 4C, Suppl. Fig. S4B). Since etiolated seedlings are not capable of doing photosynthesis, the source of carbohydrates for starch accumulation in *tpst*-mutants is likely provided by the sucrose in the growth media. Sugar uptake and transport may thus be altered in the *tpst* mutant. Indeed, the expression of the gene coding for the glucose 6-phosphate/phosphate translocator2 (GPT2), which transports glucose-6-phosphate into the plastid, is significantly upregulated 12-fold in the *tpst*-mutant (Fig. 4C, Suppl. Fig. S4B). Likewise, the gene coding for the sucrose transporter SUC1 is expressed to higher levels in *tpst*-mutants (Fig. 4C, Suppl. Fig. S4B), both consistent with increased sugar transport in *tpst*-mutants. GC-MS analysis of *tpst* seedlings showed a higher level of all saccharides and sugar phosphates, except sucrose, compared to the WT (Fig. 4D, Suppl. Table S1). These differences were not observed when sucrose was omitted from the growth medium (Suppl. Fig. S5A). Likewise, we also found a high amount of glyceric acid in *tpst* compared to the WT seedlings (Suppl. Fig. S5B), that was not observed on sucrose-free medium.

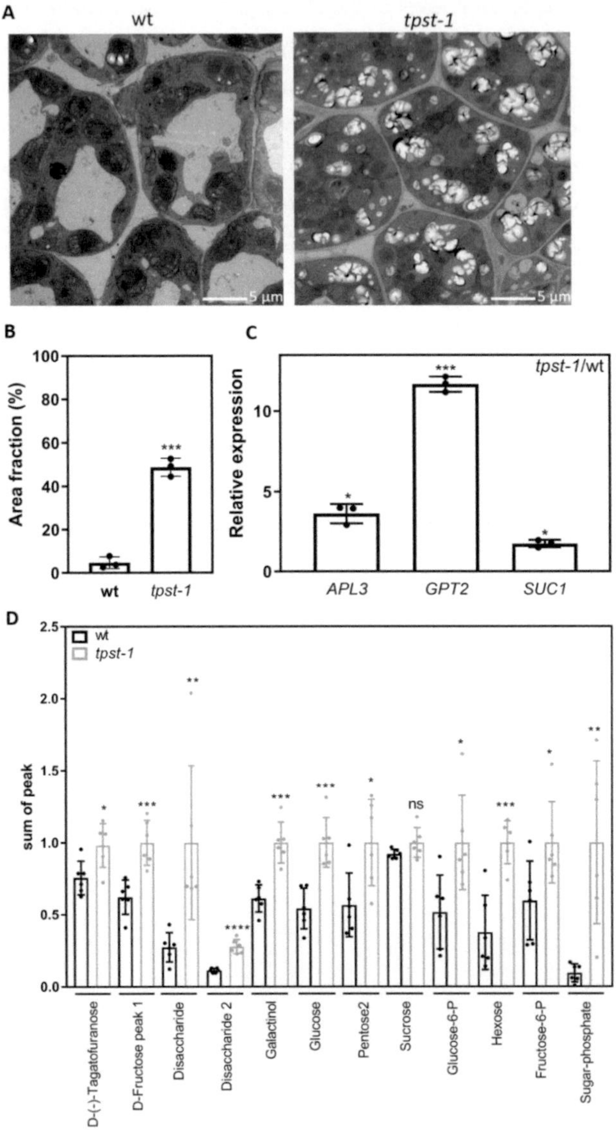

Figure 4. *tpst-1* plastid formation is disturbed by a high accumulation of starch. A Representative transmission electron microscopy images of WT and *tpst-1* etioplasts of five days etiolated seedlings. **B** Quantification of the starch accumulating in the etioplasts of five days etiolated seedlings. The amount of starch is indicated as the area fraction (%) occupied by starch grains. **C** Relative expression level of *APL3*, *GPT2* and *SUC1* in five days old

etiolated seedlings compared to the WT control. Gene Expression was normalized against two reference genes. **D** GC-MS quantification of the sugar intermediates in five days etiolated wt and *tpst-1* seedlings. Data are shown as the mean ±SD: (A, B) n=3, (C) n=3, (D) n=6. Asterisks indicate significant differences (*P<0.05, **P<0.01, ***P<0.001, ****P<0.0001; (B, D) two-tailed t-test; (C) one sample t-test).

Discussion

The TPST enzyme catalyzes the tyrosine sulfation of small posttranslationally modified peptides, which was shown to be essential for peptide activity in many cases (Whitford et al., 2012; Meng et al., 2012; Nakayama et al., 2017; Stührwohldt et al., 2020b; Okuda et al.,2020; Doll et al., 2020; Royek et al., 2022). Thereby TPST is indirectly involved in many aspects of plant growth and development including the formation of the embryonic cuticle, cell expansion, pollen tube growth, gravitropic response, plant defenses to pathogen infection and root growth and development (Stührwohldt et al., 2011; Stührwohldt et al., 2015; Zhang et al 2018; Stührwohldt et al., 2020b; Doll et al., 2020; Kaufmann et al., 2021; Onishi et al., 2022). Additionally, recent studies showed that *tpst*-mutants are hypersensitive to fructose and that TPST is important for drought stress responses (Zhong et al., 2020; Stührwohldt et al., 2021). Previous ChIP-seq analysis showed that PIF4 binds to the promoter of TPST, potentially regulating TPST expression (Oh et al., 2012). PIFs are known as negative regulators of photomorphogenesis suggesting that TPST may be involved in this process. In agreement, *tpst*-mutants were reported to display an early photomorphogenesis phenotype (Kang et al., 2014; Giorgi et al., 2021).

TPST regulates photomorphogenesis through GLK1

Here, we report that TPST activity is required to inhibit photomorphogenesis in etiolated seedlings. Mutation in TPST led to early photomorphogenesis displaying a shorter hypocotyl, open cotyledons and increased pchlide synthesis. Since GLK1 is known to regulate chlorophyll synthesis and photosynthesis-related genes (Waters et al., 2009), the high GLK1 expression level observed in etiolated *tpst* is likely causing the early photomorphogenesis phenotypes. However, the negative regulator of photomorphogenesis PIF1 was significantly more abundant in the etiolated *tpst-1* mutant. The increased production of PIF1 may be a reaction of the plant to counteract GLK1-induced early photomorphogenesis.

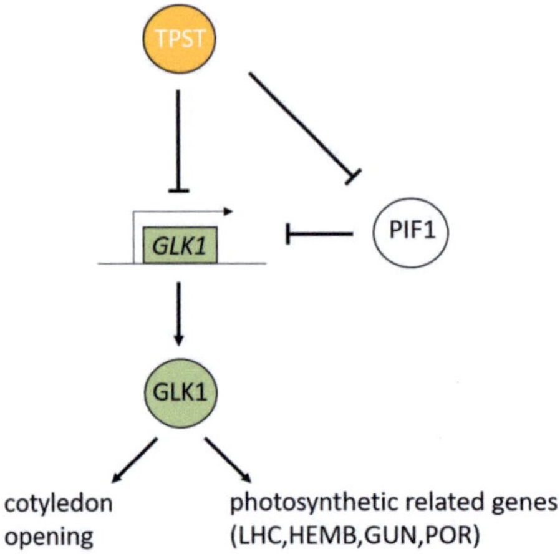

Figure 5. Simplified model of the TPST regulated photomorphogenesis. Lack of TPST leads to a higher PIF1 protein abundance. PIF1 reduces the expression of *GLK1*. Overall the lack of TPST leads to a significantly higher expression of GLK1. Thereby there must be another unknown pathway regulated by TPST which controls the *GLK1* expression. The high expression of *GLK1* leads to a higher expression of photosynthesis related genes and cotyledon opening in etiolated *tpst-1* seedlings.

Sulfated peptides are required for the photomorphogenesis

Peptides that are sulfated by TPST are known to be involved in several stress reactions, including light stress. PSK is needed for the drought stress and pathogen stress response (Stührwohldt et al., 2021; Mosher et al. 2013), the latter also involving RGF7 (Wang et al., 2021). The PSY peptide family was shown have an important role in balancing the plant growth/defence trade off (Onishi et al., 2022). RGF6 was shown to regulate the light stress induced production of anthocyanins and *RGF6* and *PSK5* were shown to be differentially expressed in response to light stress (Bühler et al., 2023). A role for RGFs and PSKs in light responses in supported by RNA-seq data of the *pifq* mutant, in which *RGF6*, *RGF9* and two *PSK* genes are differentially regulated compared to the WT (Leivar et al., 2009). Based on the light regulated expression of sulfated peptide precursors, their roles in stress responses and the early photomorphogenesis phenotypes we observed in the *tpst*-mutants, we propose an important role for sulfated peptides in photomorphogenesis. Which particular sulfated these are should be addressed in further studies.

TPST regulates the accumulation of starch in the etioplast

Differentiation of etioplasts and the etioplast-chloroplast transition are inhibited by COP1 and initiated after transfer to light (Deng et al., 1991). Etioplast structure of *tpst-1* was impaired by a high starch accumulation. Assimilatory starch is stored transiently in choroplasts, whole non-photosynthetic tissues like seeds, tubers or roots store starch in amyloplasts (Zeeman and Rees, 1999; Streb and Zeeman, 2012). Mobilizing the starch from storage is a vital process as shown by the reduced growth rate of mutants in starch mobilization, called *starch excess* (*sex*) mutants (Zeeman et al., 1998; Caspar et al., 1991). Starch overaccumulation concomitant with higher levels of sugar intermediates in the *tpst-1* mutant, is likely caused by an increased sugar uptake and transport, as suggested by a higher expression of *SUC1* and *GPT2*.

However, a high sugar uptake delivers the source to sugar transport. Thus, a high sugar transport alone cannot cause the *tpst-1* starch overaccumulation. Subsequently, the question raises how the *tpst-1* mutant takes up more sugar than the WT. Firstly, the *tpst-1* seedlings could gather more sugar through the roots. Another possibility is the uptake of sugar through defects in the *tpst-1* seedling cuticle, which were described recently (Giorgi et al., 2021). The latter seems a feasible explanation concerning the huge starch overaccumulation and should be tested in further research. However, this is not consistent with the measured GLK expression levels. Sucrose was shown to induce the GLK1 and GLK2 expression (Zhao et al., 2021). If the *tpst-1* mutant gathers more sucrose from the media than the WT, both, GLK1 and GLK2 expression levels should be accelerated. However, only GLK1 expression was increased in *tpst* mutants compared to the WT. Therefore, the induction of GLK1 expression in *tpst* seedlings additionally occurred due to another factor than sucrose levels.

Taken together, our data provide strong evidence for an important role of the TPST signal in maintaining skotomorphogenesis and for the change to photomorphogenic development. GLK1 expression is upregulated in absence of TPST, which in turn induces photomorphogenic genes like *HEMB*, *GUN5*, and *LHC*s leading to photomorphogenesis. Additionally, a putative role for the TPST in sugar transport and uptake, and starch accumulation was noticed, but needs further research and validation.

METHODS

Plant materials and growth conditions

Plant growth conditions were adapted from Bühler et al., 2023: Plant were grown on synthetic media containing 0.5 × Murashige and Skoog medium (Murashige and Skoog, 1962) supplemented with 1% sucrose and 0.38% gelrite. If indicated, 100 µM 5-aminolevulinic acid (ALA) (Sigma, St.Louis, Missouri, USA) or 0.5 mM lincomycin (Sigma, St.Louis, Missouri, USA) was added into the media. Arabidopsis seed were sterilized by a 15 minutes 70% ethanol treatment. Seeds were stratified for 2 days at 4°C. Germination was induced by a 4h-6h light treatment (120 µmol m^{-2} s^{-1} white light). Afterwards plants were grown at 22° C in darkness or under short-day (12h/12h) conditions. For greening rate experiments, seedlings were grown for three to six days in the dark and then transferred into constant light (120 µmol m^{-2} s^{-1} white light) for two days.

Arabidopsis lines

All transgenic lines used for this study were previously described; *tpst-1 (Komori et al., 2009), tpst-2 (Kang et al., 2014).*

Quantification of cotyledon opening

Seedlings were grown for three-day under weak light (1 µmol m^{-2} s^{-1}) with or without 0,5 mM lincomycin in the media. The cotyledons were counted as open if the cotyledon were separated from each other.

Pchlide extraction and determination

Exactly thirty etiolated seedlings were harvested under green safe light and frozen in liquid nitrogen. They were homogenized in a TissueLizer LT (Qiagen, Venlo, Netherlands) and dissolved in ice-cold 80% acetone. After shaking for 5 h in darkness the probes were centrifuged at 13000 g for 10 min and the supernatant was transferred to a new tube. To determine the pchlide amount 100 µl were transferred in a 96 well microplate (flat bottom, greiner bio-one, Kremsmünster, Austria). The fluorescence was excited at 440 nm and scanned from 600 to 750 nm by a fluorescence spectrophotometer (TECAN Spark plate reader, https://lifesciences.tecan.com/multimode-plate-reader).

qPCR

RNA extraction and qPCR protocols were used as described in Bühler et al., 2023: Total RNA was extracted using TRIzol (ABP Biosciences, Beltsville, Maryland, USA) according to the manufactors protocol. Up to 5µg RNA was used to produce cDNA by RevertAid reverse transcriptase (Thermo Fisher Scientific, Waltham, Massachusetts, USA) and oligo(dT) primers. qPCR primers are listed in supplemental table 1. qPCR was done in a CFX96 Real-Time PCR Detection system (BioRad, Hercules, California, USA). Experiments were performed in three biological and two technical replicates. Expression data were normalized against the two housekeeping genes *ACTIN2* (*At3g18780*) and *TUB4* (*At5g44340*)

Transmission electron microscopy

The cotyledons from five-day etiolated seedlings were cut from the stem and once in a transverse plane. Afterwards, they were transferred into the fixative (2.5% glutaraldehyde in 0,1 M Sorensen phosphate buffer pH 7.0 (NaH_2PO_4/Na_2HPO_4) and incubated in vacuum chamber for 20 min. Pressure was slowly reduced to 800 mbar. After vacuum infiltration samples were kept in the fixative at room temperature overnight. Samples were mounted on TEM grids and contrasted with uranyl acetate und lead citrate. Images were taken over a whole leaf at 4,900x magnification. Followed by a SUR (Systematic Uniform Random) sampling principle. Magnification was selected as the lowest magnification at which we could reliably identify the structures of interest. For the quantification of starch in the etioplasts a grid of was placed over the images in Image J. One square (outlined by four crosses at the corners) is 0.50 μm^2. Partial counts made on individual images were put together to make total counts. Area fraction of starch in etioplasts was calculated from total counts.

Protein extraction and immunoblotting

Whole five-day etiolated seedlings were harvested under green safety light and immediately frozen in liquid nitrogen. Samples were homogenized using a TissueLizer LT (Qiagen, Venlo, Netherlands). Total protein was extracted in protein extraction buffer (500 mM NaCl, 250 mM Tris/HCl pH 7.5, 2.5 % Triton X100, 50 mM ß-mercaptoethanol). Afterwards samples were incubated for 10 min at 95 °C with SDS loading dye. Samples were separated by SDS-PAGE and blotted on a nitrocellulose membrane using a semi dry blotter (BioRad, Hercules, California, USA). The membrane was blocked in 5 % skim milk powder, 0.1 % Tween 20. For the detection of PIF1, a specific PIF1 Antibody (1:1000, PhytoAB, San Jose, California, USA) and a secondary goat anti-rabbit antibody horseradish peroxidase (HRP) conjugate (1:10000, EMD Milipore, Burlington, Massachusetts, USA) was used. For the detection of PIF3 a specific PIF3 antibody (1:1000, Agrisera, Vännäs, Sweden) and a secondary anti-goat antibody HRP conjugate (1:10000, Invitrogen, Waltham, Massachusetts, USA) was used. Blots were developed with SuperSignal™ west dura extended duration substrate (Thermo Fisher Scientific, Waltham, Massachusetts, USA) chemiluminescence kit. Li-Cor Odyssee imager (Li-Cor Biosciences, Bad Homburg, Germany) was used to detect the chemiluminescence.

GC-MS

Protocol was adapted from (Krawczyk 2021): Primary metabolite and sterol extraction were performed as previously described (Rotsch et al., 2017). Ten seedlings were collected per sample. After tissue disruption with a shaking mill and glass beads, 250 µL extraction solution (1:1 v/v methanol/chloroform) were added per sample, vortexed and incubated for 30 minutes at room temperature under constant shaking. Then, 0.00125 mg *allo*-inositol in 250 µL water was added as an internal standard and vortexed. Samples were centrifuged 2 minutes at full speed in a microfuge and the aqueous phase containing the metabolites was transferred to a fresh tube. 200 µL were evaporated under N_2 stream and used for derivatisation with 15 µL of methoxyamine hydrochloride in pyridine (30 mg/mL) overnight at room temperature. Derivatisation with 30 µL *N*-methyl-*N*-(trimethylsilyl) trifluoroacetamide (MSTFA) followed for at least 1 h to obtain methoxyimino (MEOX)- and trimethylsilyl (TMS)-derivatives of the metabolites (Bellaire et al., 2014).

GC-MS measurements were performed as previously described (Touraine et al., 2019). The spectra were identified with the Golm metabolome database (GMD) and the National Institute of Standards and Technology (NIST) spectral library 2.0f. Chemical information on metabolites identified within the GMD can be obtained at http://gmd.mpimp- golm.mpg.de/search.aspx (Kopka *et al.*, 2005). Due to the high levels, sucrose was measured in separate runs a higher split. Data were analysed using MSD ChemStation (F.01.03.2357). Masses used for quantification are depicted in Supplemental Table. 2.

Statistical analysis

Statistic analysis was done using GraphPad Prism (Dotmatics, Boston, USA). To determine significant differences the following tests were performed. One sample *t*-tests against the expression ratio of 1 were performed for q-PCR evaluation. Two tailed *t*-test were performed for cotyledon opening, greening rate, pchlide measurement, protein abundance level (grey value), starch abundance and GC-MS experiments.

Supplemental Information

Supplemental Table 1: qPCR primer.

Supplemental Table 2: Relative abundances of metabolites of etiolated wild-type and *tpst-1* seedlings in the presence or absence of sucrose.

Acknowledgements

The authors thank Lhana Stein for excellent technical support. The authors thank Ryota Komori for the *tpst-1* line (Komori et al., 2009) and Jun Kang for the *tpst-2* line (Kang et al., 2014).

Supplemental Figures

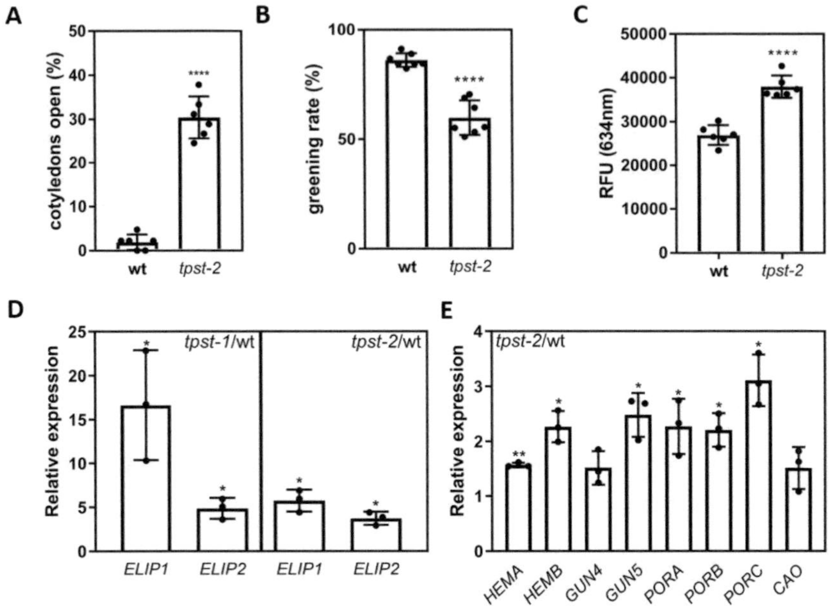

Supplemental Figure S1. *tpst-2* shows the same early photmorphogenesis phenotypes as *tpst-1*. A Determination of the open cotyledons of five days etiolated seedlings. **B** Determination of green cotyledons. Percentage of green cotyledons is shown for five-day etiolated seedlings after transfer to two days of constant light. **C** Pchlide measurement of five-day etiolated seedlings. The concentration of pchlide is shown in relative fluorescence units (RFU) at 634 nm. **D** Relative expression level of *ELIP1* and *ELIP2* in five-day etiolated *tpst-1* or *tpst-2* seedlings compared to the WT control. Gene Expression was normalized against two reference genes. **E** Relative expression level of *HEMA*, *HEMB*, *GUN4*, *GUN5*, *PORA*, *PORB*, *PORC* and *CAO* in five-day etiolated seedlings compared to the WT control. Gene Expression was normalized against two reference genes (*ACTIN2* and *TUB4*). Data are shown as the mean ±SD: (A) n=6, (B) n=7, (C) n=6, (D, E) n=3. Asterisks indicate significant differences (*P<0.05, **P<0.01, ****P<0.0001; (A, B, C) two-tailed t-test; (D, E) one sample t-test).

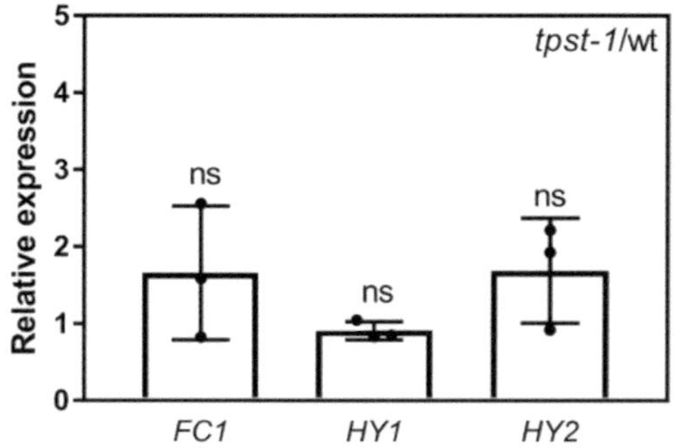

Supplemental Figure S2. Phytochromibilin synthesis is not altered in the *tpst-1* mutant. A Relative expression level of *FC1*, *HY1* and *HY2* in five days old etiolated seedlings compared to the WT control. Gene Expression was normalized against two reference genes (*ACTIN2* and *TUB4*). Data are shown as the mean ±SD: (A) n=3. ns indicate no significant differences (one sample t-test).

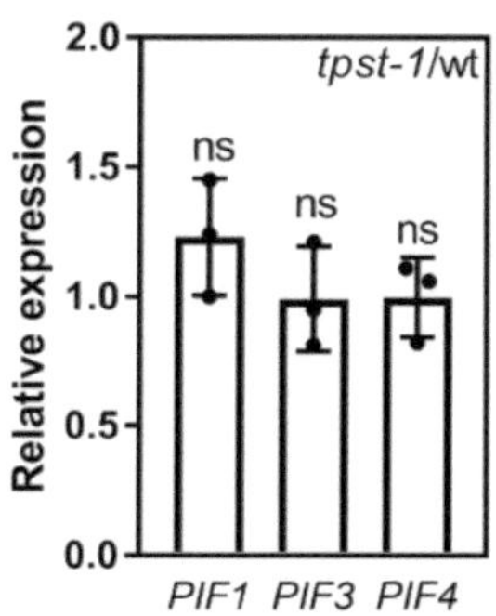

Supplemental Figure S3. PIF1, PIF3 and PIF4 expression is not altered in the *tpst-1* mutant. A Relative expression level of *PIF1*, *PIF3* and *PIF4* in five days old etiolated seedlings compared to the WT control. Gene Expression was normalized against two reference genes (*ACTIN2* and *TUB4*). n=3. Data are shown as the mean ±SD. ns indicates no significant difference (one sample t-test).

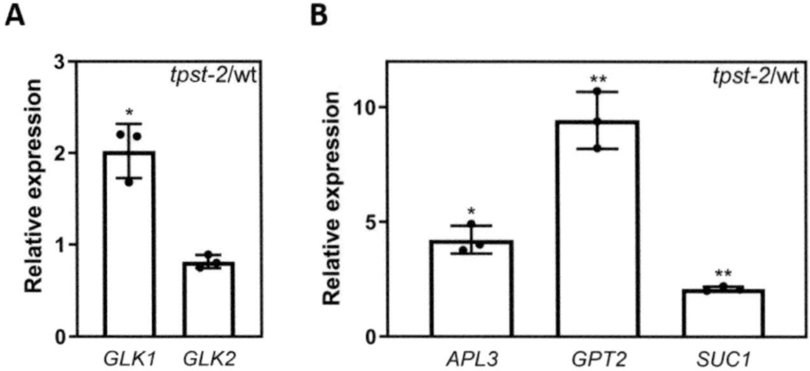

Supplemental Figure S4. A Relative expression level of *GLK1* and *GLK2* in five-day etiolated seedlings compared to the WT control. Gene Expression was normalized against two reference genes. **B** Relative expression level of *APL3*, *GPT2* and *SUC1* in five-day etiolated seedlings compared to the WT control. Gene Expression was normalized against two reference genes. Data are shown as the mean ±SD: (A-C) n=3. Asterisks indicate significant differences (*P<0.05, **P<0.01; one sample t-test).

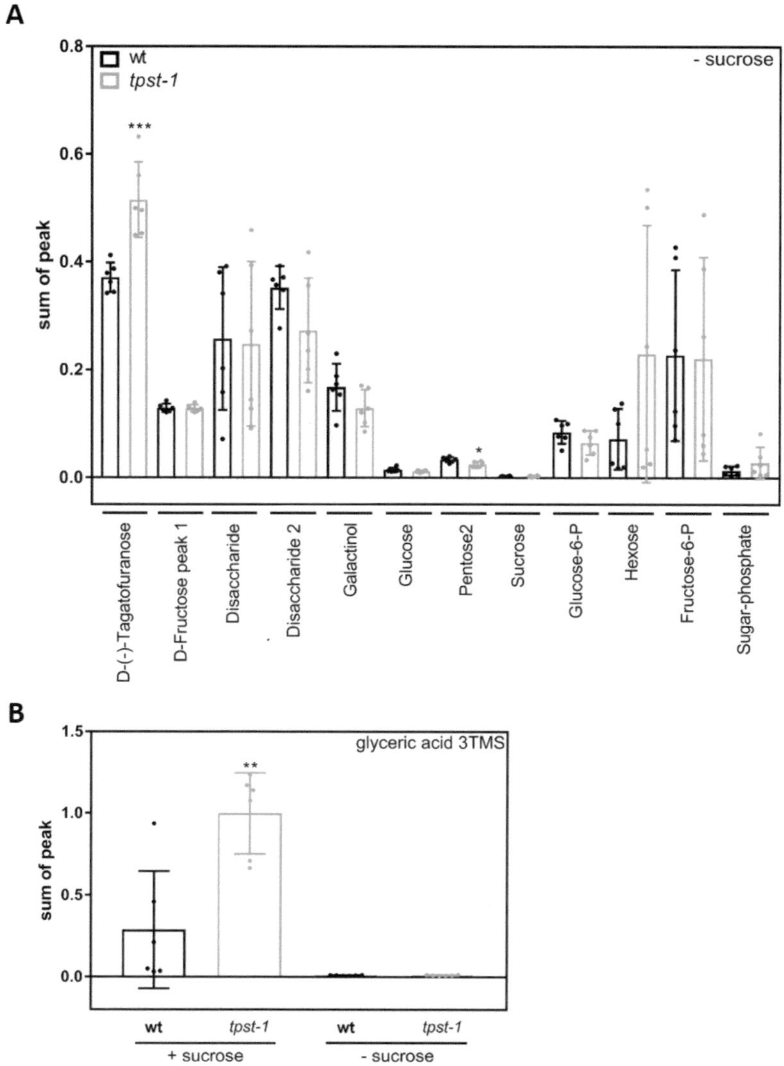

Supplemental Figure S5. Glyceric acid amount in *tpst-1* is highly increased. A GC-MS quantification of the sugar intermediates in five days etiolated wt and *tpst-1* seedlings without sucrose in the media. **B** GC-MS quantification of the glyceric acid amount in five days etiolated wt and *tpst-1* seedlings in the precence or abcence of sucrose. Data are shown as the mean ±SD: (A) n=6, (B) n=6. Asterisks indicate significant differences (**P<0.01, ***P<0.001; two-tailed t-test).

Supplemental Table S1. qPCR Primer.

Accession No.	Name	Sequence (5´-3´)	Publication
At3g18780	*ACTIN2* RT F	GTGGATATCAGGAAGGATCTGTACGGT	Bühler et al., 2023
At3g18780	*ACTIN2* RT R	TTCTGTGAACGATTCCTGGACCTGC	Bühler et al., 2023
At5g44340	*Tub4* RT F	TGAGGAAGGAAGCTGAGAACAG	Bühler et al., 2023
At5g44340	*Tub4* RT F	ATACACTGGTCAGCGTTTTC	Bühler et al., 2023
At1g58290	*HEMA* RT F	GTTGCTGCCAACAAAGAAGA	
At1g58290	*HEMA* RT R	AATCCCTCCATGCTTCAAAC	
At1g69740	*HEMB* RT F	TCTGATGGTCACGATGGGATAG	
At1g69740	*HEMB* RT R	GCTGAGCGGATTGCACCTACT	
At5g13630	*GUN5* RT F	CAGCCAACATCAGTCTTGCT	
At5g13630	*GUN5* RT R	ACCTGCTTCTTCTCAGCCAT	
At3g59400	*GUN4* RT-F	ACAAACGCCTTGAGAGGAACGC	
At3g59400	*GUN4* RT-R	TATCATCAGCTGTCGCAAACGC	
At2g20180	*PIF1* RT-F	ACTTATACCTCGCTGCAACAAGTC	
At2g20180	*PIF1* RT-R	CATGCCCGGATACATCATTGGC	
At1g09530	*PIF3* RT-F	GGGTTTGGGTTCAAAGAGAAGCC	
At1g09530	*PIF3* RT-R	TTGATCCTATCACGCCGTCTCC	
AT3G22840	*ELIP1* RT F	TGAGGGAGGACCCACGAATGAA	Sun et al., 2019
AT3G22840	*ELIP1* RT R	CACCGTCGGAGATCTGAGCTAAAAC	Sun et al., 2019
AT4G14690	*ELIP2* RT F	GCTCAGGGCGATCCTATCAAGGAAG	Sun et al., 2019
AT4G14690	*ELIP2* RT R	CCACCGTCAGAGATCTGAGCAAACA	Sun et al., 2019
AT2G20570	*GLK1* RT F	TTCTACCGCCATGCCTAATCCG	
AT2G20570	*GLK1* RT R	AACTGGCGGTGCTCTAAATCTCG	
AT5G44190	*GLK2* RT F	TCAAGCATCGGTGTTCCCACAAG	
AT5G44190	*GLK2* RT R	TCGAGGGATGAATGTCGATGGG	
At1g44446	*CAO* RT F	ATCGAAAGCACACCAACAGGTTC	
At1g44446	*CAO* RT R	TTCTTCCATGTGGGCTAACTTGTC	
AT4G39210	*APL3* RT F	CACACGGATGTTTCTTGGGA	Zhang et al., 2020
AT4G39210	*APL3* RT R	GGTAACTATCCGCTCCTAAC	Zhang et al., 2020
AT1G71880	*SUC1* RT F	GATACCACAAATGATAGTCTCACTAG	Shin et al., 2012
AT1G71880	*SUC1* RT R	GTTGCCTTGGGTGCGTC	Shin et al., 2012
AT1G61800	*GPT2* RT F	TTTCTATTGCCGTGGAAGGTCCTC	
AT1G61800	*GPT2* RT R	ACTTTGTGCCACTACCCACCAG	
AT3G54890	*LHCA1* RT F	AGATGGGCTATGCTCGCTGTTC	
AT3G54890	*LHCA1* RT R	ATGGCCAAGATTGTGGGCAAAG	
AT3G61470	*LHCA2* RT F	AGTTGACTGTTCCAGCTGTGTCC	
AT3G61470	*LHCA2* RT R	TAGACTGTCCGGGTCCGATGAAAG	
AT2G05100	*LHCB2.1* RT F	TACTCCTCAAAGCATCTGGTACGG	
AT2G05100	*LHCB2.1* RT R	AGCACCCAACATAGCCCATCTG	
AT1G29920	*LHCB1.1* RT F	AGGAACGGAGTCAAGTTTGGAGAG	
AT1G29920	*LHCB1.1* RT R	CCATTTCCTGCGACTCTGTAACCT	

Supplemental table S2. Relative abundances of metabolites of etiolated wild-type and *tpst-1* seedlings in the presence or absence of sucrose.

	averaged					standard deviation			
	wt + sucrose	*tpst-1* + sucrose	wt	*tpst-1*		wt + sucrose	*tpst-1* + sucrose	wt	*tpst-1*
amino acids									
*alanine (2TMS)	0.45	1.00	0.02	0.07		0.23	0.54	0.01	0.15
*aspartic acid (3TMS)	1.00	0.63	0.17	0.04		0.68	0.50	0.04	0.01
*beta-Alanine 3TMS	0.16	0.13	0.79	0.77		0.15	0.09	0.19	0.24
*GABA (3TMS)	0.27	0.28	0.32	0.15		0.17	0.25	0.08	0.07
*glutamic acid (3TMS)	0.81	0.90	0.19	0.05		0.48	0.94	0.07	0.02
*glycine (3TMS)	0.55	0.60	0.92	0.56		0.23	0.33	0.16	0.24
*isoleucine (2TMS)	0.17	0.14	1.00	0.38		0.10	0.03	0.40	0.24
*Leucine (2TMS)	0.17	0.11	1.00	0.43		0.16	0.04	0.42	0.30
*L-Glutamine, 3TMS	0.73	1.00	0.05	0.00		0.79	1.50	0.04	0.00
*Phenylalanine 2TMS	0.09	0.06	1.00	0.28		0.04	0.04	0.23	0.11
*Proline 2TMS	0.10	0.10	0.99	0.56		0.11	0.04	0.54	0.45
*Pyroglutamic acid 2TMS	0.80	1.00	0.09	0.02		0.20	0.13	0.01	0.02
*serine 3TMS	0.17	0.07	0.67	0.38		0.33	0.04	0.21	0.22
*Threonine 2 TMS	0.36	0.46	0.51	0.20		0.11	0.13	0.08	0.08
*Valine 1TMS	0.27	0.24	1.00	0.52		0.08	0.06	0.15	0.15
organic acids									
*citric acid (4TMS)	0.67	0.89	0.19	0.11		0.27	0.16	0.04	0.05
*fumarate	0.40	1.00	0.37	0.23		0.09	0.27	0.06	0.11
*Glyceric acid 3TMS	0.24	1.00	0.01	0.01		0.32	0.25	0.00	0.00
*Glycolic acid 2TMS	0.43	0.51	0.64	0.62		0.04	0.08	0.15	0.26
*Lactate	0.68	0.64	0.59	0.63		0.17	0.12	0.53	0.62
*malic acid (3TMS)	0.78	1.00	0.07	0.04		0.15	0.20	0.02	0.03
*Malonic acid 2TMS	0.04	0.09	1.00	1.00		0.01	0.01	0.18	0.41
*phosphoric acid (3TMS)	1.00	0.91	0.96	0.70		0.19	0.15	0.21	0.20
*succinic acid (2TMS)	0.52	0.73	0.61	0.60		0.09	0.17	0.11	0.29
*Threonic Acid	0.64	1.00	0.02	0.02		0.17	0.19	0.00	0.01
sugars									
*D-(-)-Tagatofuranose	0.79	0.98	0.37	0.52		0.12	0.15	0.03	0.07
*D-Fructose peak 1	0.62	1.00	0.13	0.13		0.10	0.16	0.01	0.01
*disaccharide	0.39	1.00	0.26	0.25		0.25	0.53	0.13	0.15
*disaccharide 2	0.12	0.28	0.35	0.27		0.02	0.05	0.04	0.10
*Galactinol	0.69	1.00	0.17	0.13		0.16	0.14	0.04	0.03
*glucose (1MEOX,5TMS) Peak1	0.56	1.00	0.01	0.01		0.13	0.17	0.00	0.00
*pentose2	0.60	1.00	0.03	0.02		0.20	0.30	0.00	0.01
*sucrose	0.93	1.00	0.00	0.00		0.03	0.10	0.00	0.00
Glucose-6-P	0.72	1.00	0.08	0.07		0.44	0.33	0.02	0.02
hexose	0.40	1.00	0.07	0.23		0.23	0.15	0.06	0.24
fructose-6-P	0.82	1.00	0.23	0.22		0.48	0.28	0.16	0.19
sugar-phosphate	0.14	1.00	0.01	0.03		0.11	0.56	0.01	0.03
polyols									
*1,3-Propanediol 2TMS	0.97	1.00	0.69	0.61		0.08	0.06	0.11	0.12
*glycerol (3TMS)	1.00	0.96	0.06	0.07		0.08	0.11	0.02	0.05
*Inositol-P	0.95	1.00	0.49	0.49		0.37	0.16	0.14	0.15
*myo-Inositol	0.59	1.00	0.40	0.41		0.10	0.06	0.05	0.09
others									
*DL-Ornithine 4TMS	0.77	0.35	1.00	0.25		0.28	0.33	0.47	0.10
*Ethanolamine 3TMS	0.18	0.11	0.55	0.37		0.14	0.08	0.08	0.09
*Hydroxylamine 3TMS	1.00	0.72	0.15	0.17		0.38	0.22	0.05	0.06
*Phosphorylethanolamine, 4TMS	0.16	0.16	0.41	0.33		0.09	0.06	0.12	0.15
*Putrescine	0.34	0.35	0.53	0.25		0.14	0.30	0.13	0.10
*Pyrogallol 3TMS	0.36	0.41	1.00	0.75		0.05	0.06	0.17	0.08
unknowns									
*U1	1.00	1.00	0.25	0.02		0.29	0.78	0.32	0.01
*U2	0.73	0.88	0.48	0.51		0.11	0.14	0.26	0.11
*U3	0.38	1.00	0.04	0.01		0.23	0.31	0.07	0.01

Quantification was done as described in Touraine et al., 2019: The values quantified for a single ion were divided by the fresh weight of the sample and the value of the internal standard allo - inositol. When substances were represented by two or more analytes, these were added based on their total ion count. Then, the averages were calculated and the highest value set to 1.

2.3 Phytosulfokine (PSK) precursor processing by subtilase SBT3.8 and PSK signaling improve drought stress tolerance in Arabidopsis

Nils Stührwohldt[1,#], Eric Bühler[1], Margret Sauter[2], and Andreas Schaller[1]

[1] Department of Plant Physiology and Biochemistry, Institute of Biology, University of Hohenheim, 70593 Stuttgart, Germany.

[2] Plant Developmental Biology and Physiology, University of Kiel, 24118 Kiel, Germany

[#]correspondence: nils.stuehrwohldt@uni-hohenheim.de

Status: published

Original publication: Journal of Experimental Botany, Volume 72, Issue 9, 13 April 2021, Pages 3427–3440.
By permission of the Society for Experimental Biology

Publication available at:
https://academic.oup.com/jxb/article/72/9/3427/6104858?login=false

Own contribution: I performed the experiments shown in figure 5C, figure 6A and B and figure 7C, D and E.

Abstract

Increasing drought stress poses a severe threat to agricultural productivity. Plants, however, have evolved numerous mechanisms to cope with such environmental stress. Here we report that the stress-induced production of a peptide signal contributes to stress tolerance. The expression of phytosulfokine (PSK) peptide precursor genes, and transcripts of three subtilisin-like serine proteases, SBT1.4, SBT3.7, and SBT3.8, were found to be up-regulated in response to osmotic stress. Stress symptoms were more pronounced in *sbt3.8* loss-of-function mutants and could be alleviated by PSK treatment. Osmotic stress tolerance was improved in plants overexpressing the PSK1 precursor (proPSK1) or SBT3.8, resulting in higher fresh weight and improved lateral root development in transgenic plants compared with wild-type plants. We further showed that SBT3.8 is involved in the biogenesis of the bioactive PSK peptide. ProPSK1 was cleaved by SBT3.8 at the C-terminus of the PSK pentapeptide. Processing by SBT3.8 depended on the aspartic acid residue directly following the cleavage site. ProPSK1 processing was impaired in the *sbt3.8* mutant. The data suggest that increased expression of proPSK1 in response to osmotic stress followed by the post-translational processing of proPSK1 by SBT3.8 leads to the production of PSK as a peptide signal for stress mitigation.

Introduction

Plants have evolved mechanisms to adapt to their environment and are thus able to survive unfavorable stress conditions. Due to climate change, the ability to cope with increasing temperatures and water scarcity is becoming more and more important. Plant responses to these abiotic stress factors are regulated at multiple levels, including through peptide signals (Takahashi and Shinozaki, 2019).

Several classes of plant signaling peptides are distinguished based on differences in their structure and biogenesis. Some peptides are derived from larger precursor proteins that may or may not have a function on their own. Other peptides are produced from small peptide-encoding genes, or translated from short open-reading frames within the 5′-untranslated regions of protein-coding genes, or from the primary transcript of miRNAs (Tavormina et al., 2015). Precursor-derived peptides can be divided into cysteine-rich peptides that depend on disulfide bond formation for activity, and the large group of post-translationally modified peptides (Matsubayashi, 2014). The latter require proteolytic processing to release the peptide moiety from the inactive precursor, and may rely on tyrosine sulfation, proline hydroxylation, and arabinosylation of hydroxyprolines for full activity (Stührwohldt and Schaller, 2019).

Recent publications have reported that processing of peptide precursors is essential for the biogenesis of mature and active peptides, and that the processing activity may be the limiting factor for peptide signaling. IDA (Inflorescence Deficient in Abscission) is processed by subtilisin-like serine proteases (SBTs) SBT4.12, SBT4.13, and SBT5.2, and plants lacking SBT activity are impaired in floral organ abscission (Schardon et al., 2016). The tyrosine-sulfated CLEL6 and CLEL9 peptides are activated by several steps of proteolytic processing that occur in consecutive compartments of the secretory pathway (Stührwohldt et al., 2020b). Following cleavage of the signal peptide upon entry into the endoplasmic reticulum, CLEL6 and CLEL9 are pre-processed by SBT6.1 in the cis-Golgi and activated by aspartate (Asp, D)-dependent SBTs including SBT3.8 in post-Golgi compartments (Ghorbani et al., 2016; Stührwohldt et al., 2020b). Processing and release of the PEP1 peptide as an immune-modulatory phytocytokine are regulated by calcium-dependent activation of metacaspase 4 (Hander et al., 2019). The formation of the embryonic cuticle is controlled by the SBT2.4 (ALE1)-dependent activation of

Twisted Seed 1 (TWS1) (Doll et al., 2020), and the drought-induced formation of phytosulfokine (PSK) by phytaspase 2 regulates flower drop in tomato (Reichardt et al., 2020).

In addition to drought-induced tomato flower drop, the disulfated PSK pentapeptide [Y(SO3H)-I-Y(SO3H)-T-Q] is also involved in the regulation of several developmental processes and both biotic and abiotic interactions. Briefly, PSK promotes growth of roots and hypocotyls mainly by the regulation of cell expansion by a plasma-membrane-localized module including the leucine-rich repeat receptor kinases PSKR1 and PSKR2, the co-receptor BRI1-associated receptor kinase 1 (BAK1), the proton pump AHA1, and the cyclic nucleotide-gated channel CNGC17 (Matsubayashi et al., 2006; Kutschmar et al., 2009; Stührwohldt et al., 2011; Ladwig et al., 2015). Furthermore, in *Zinnia elegans*, PSK promotes the differentiation of mesophyll cells into tracheary elements by down-regulation of stress-response genes such as chitinases, redox-related enzymes, or ethylene synthesis genes at the onset of the transdifferentiation process (Matsubayashi et al., 1999; Motose et al., 2009). Defense responses against biotrophic and necrotrophic pathogens are affected differentially by PSK signaling. Arabidopsis plants deficient in PSK receptor genes are more susceptible to pathogens with a necrotrophic lifestyle, whereas resistance against hemibiotrophs or biotrophs is increased (Loivamäki et al., 2010; Igarashi et al., 2012; Mosher et al., 2013; Rodiuc et al., 2016). Also in tomato, loss of PSK receptor function led to increased susceptibility to the necrotroph *Botrytis cinerea* (Zhang et al., 2018). In contrast, PSK attenuates pattern-triggered immunity in Arabidopsis, and the PSK receptor mutant *pskr1* shows increased resistance against *Pseudomonas syringae* (Igarashi et al., 2012).

PSK depends on tyrosine sulfation for bioactivity. Tyrosine sulfation is driven by Golgi-localized tyrosylprotein sulfotransferase (TPST), which is encoded by a single-copy gene in Arabidopsis (Komori et al., 2009). The enzymes that are responsible for precursor processing to release the PSK peptide are essentially unknown in Arabidopsis. SBT1.1 was shown to cleave Arabidopsis PSK4 precursor peptides *in vitro* (Srivastava et al., 2008), but the cleavage site is upstream of the mature peptide within the variable part of the precursor, and the physiological relevance of this cleavage and of SBT1.1 for PSK biogenesis remain unclear. In tomato, an Asp-specific SBT (tomato phytaspase 2, SlPhyt2) was shown to be involved in PSK maturation (Reichardt et al., 2020). Indeed, all eight PSK precursors in tomato (Reichardt et al., 2020) and the seven PSK precursors in Arabidopsis (Kaufmann and Sauter, 2019) share an Asp residue on the amino side of the PSK pentapeptide. Arabidopsis proPSK1 is the only precursor featuring Asp at both the N-terminal and the C-terminal processing sites. However, all PSK precursor genes share the conserved sequence of the mature peptide (Y-I-Y-T-Q). Therefore, in contrast to all other known peptide families in plants, all PSK precursors give rise to the same bioactive pentapeptide, Y(SO3H)-I-Y(SO3H)-T-Q (Fig. 1A), which was first isolated from asparagus cell cultures (Matsubayashi and Sakagami, 1996). The only exception is PSK6, which differs in the last amino acid of the PSK pentapeptide. However, since no expressed sequence tags have been reported and expression levels, according to RNA-seq data, are extremely low, PSK6 is considered a pseudogene (Kaufmann and Sauter, 2017).

Inspired by the finding that PSK mediates drought-induced flower drop in tomato (Reichardt et al., 2020), we investigated whether PSK and SBT-mediated PSK maturation are also involved in drought stress responses in Arabidopsis. Because drought stress is difficult to standardize under experimental conditions, mannitol-induced water scarcity is commonly used as an experimental approximation of drought (Hazman et al., 2016). In this work, we identified PSK genes and SBTs that are up-regulated upon mannitol treatment. We identify SBT3.8 from Arabidopsis as the limiting factor in the osmotic stress response and the formation of PSK from its propeptide. We describe a novel mechanism of peptide-dependent adaptation to osmotic stress that relies on SBT3.8-mediated release of PSK as the stress signal.

Material and Methods

Plant materials and growth conditions

For growth experiments on synthetic media, Arabidopsis seeds were surface-sterilized in 70% (v/v) ethanol for 15 minutes, washed in 100% ethanol, and laid out on square plates containing 0.5 × Murashige and Skoog medium (Murashige and Skoog, 1962), 1% (w/v) sucrose, and 0.38% (w/v) gelrite. Seeds were stratified for 1 day at 4° C and grown under short-day conditions (12 h photoperiod) at 22 °C and 100–120 µmol m^{-2} s^{-1} white light. For osmotic stress treatment, 5-day-old seedlings were transferred to plates containing the same medium supplemented with 300 mM mannitol. Growth phenotypes were scored after 6 days and compared for plants grown on control or mannitol-containing media. Root and hypocotyl length were measured using ImageJ (http://rsbweb.nih.gov/ij/). All experiments were carried out at least three times; representative results are shown. The disulfated PSK pentapeptide (Tyr(SO3H)-Ile-Tyr(SO3H)-Thr-Gln) was added at 1 µM. The disulfated pentapeptide was synthesized by PepMIC (Souzhou, China) and obtained at >95% purity. Lyophilized peptides were reconstituted in water. The peptide concentration was determined by a modified Lowry procedure using the DC Protein Assay (BioRad; Munich, Germany) according to the manufacturer's instructions. Seeds of tobacco (*Nicotiana benthamiana*) were obtained from Agroscience (Agroscience GmbH, Neustadt, Germany) and were grown under a day length of 16 h at 28 °C. Tobacco plants were fertilized 3 times a week with universal fertilizer (Wuxal®, 2 ml l^{-1}) and used for experiments at the age of 4–6 weeks. The *sbt* loss-of-function mutants and the SBT3.8ox transgenic lines have been described before (Rautengarten et al., 2005; Stührwohldt et al., 2020b). For experiments on soil, Arabidopsis plants were grown on potting compost with 3.6 % (v/v) sand and 7.2 % (v/v) perlite. To minimize the risk of thrips infection, seeds were incubated overnight at –75 °C. They were resuspended in 0.1% (w/v) agar, stratified for 48 h at 4 °C in the dark, and grown under short-day conditions (12 h photoperiod) at 22 °C and 100–120 µmol m^{-2} s^{-1} white light. Plants were watered every 2 days. For drought stress experiments, wild-type and SBT3.8ox plants were grown for 3 weeks with sufficient water. Subsequently, plants were watered only every third day with 10 ml water each.

Generation of expression constructs

All primers used for amplification of constructs by PCR are listed in Supplementary Table S1. ORFs of PSK1 and RGF1 lacking the predicted signal peptides were amplified by PCR from cDNA and cloned into the NcoI restriction site of pETDuet1 (Novagen/Merck KGaA, Darmstadt, Germany). C-terminal His tags were added by including six His codons in the reverse PCR primers. Correct orientation and translational fusion with the C-terminal His-tag were verified by sequencing. PSK1 and RGF1 D/A mutants were created by site-directed mutagenesis or PCR with a reverse primer including mutations, respectively. For the PSK1 multiple D/A mutant, D8A, D28A, D30A, and D39A mutations were inserted step by step by site-directed mutagenesis. All mutations were confirmed by sequencing.

Expression and purification of recombinant proteins

Escherichia coli cells were grown in Luria–Bertani medium with appropriate antibiotics at 220 rpm and 37 °C to an A600 of 0.6. Protein expression in *E. coli* BL21 was induced by 1 mM isopropyl β-D-1-thiogalactopyranoside for 2 hours at 30 °C. His-tagged PSK1 or RGF1 precursors were purified from bacterial extracts by metal chelate affinity chromatography on Ni-NTA agarose (Qiagen, Hilden, Germany) according to the manufacturer's recommendations. Cells were harvested by centrifugation (4000 g, 4 °C, 20 min) and resuspended in 50 mM sodium phosphate buffer, pH 7.0, supplemented with 300 mM NaCl including lysozyme, phenylmethylsulfonyl fluoride, DNase, and Proteinase Inhibitor Cocktail (Serva). Cells were lysed by sonication (SONOPULS HB2070 with MS72 sonic needle, Bandelin Electronics, Berlin, Germany) three times at 4 °C for 30 s. The soluble fraction was collected by centrifugation (15 000 g, 4 °C, 10 min). Recombinant proteins were bound to the Ni-NTA column for at least 1 h. The column was washed extensively in buffer containing 10 mM imidazole and recombinant proteins were eluted in buffer containing 200 mM imidazole. After ultrafiltration (30 kDa molecular mass cutoff), the recombinant protein in the filtrate was further purified by size exclusion chromatography (NGC chromatography system with Enrich SEC 650 column, BioRad, Munich, Germany) in 50 mM NaH_2PO_4/Na_2HPO_4, pH 5.5 or pH 7.0, respectively, and 10 mM NaCl.

Generation of PSK1 overexpressors

The cDNAs of rice (*Oryza sativa*) PSK1 and the site-directed D79A, D85A double mutant were a gift of Andrey Vartapetian (Moscow State University). They were obtained in translational fusion with enhanced green fluorescent protein between the CaMV 35S promoter and terminator in pRTL2 (Carrington et al., 1990). The entire expression cassettes were cut out from pRTL2 with PstI and cloned into pGreenII 0229. Plasmids were transformed into *Agrobacterium tumefaciens* GV3101 and transformed into *Arabidopsis thaliana* Col-0 by floral dip (Clough and Bent, 1998). Transgenic lines were selected on glufosinate, and homozygous lines in the T3 or T4 generation were used in further experiments.

Transient expression in *N. benthamiana* and protein extraction

Agrobacterium tumefaciens strains C58C1 and GV3101 were used for transient expression experiments in *N. benthamiana*. Bacteria were grown on plates containing appropriate antibiotics (rifampicin, tetracycline, and spectinomycin for C58C1, and gentamycin and spectinomycin for GV3101) at 28 °C and were washed off the plates in 10 mM MES, pH 5.6, containing 10 mM $MgCl_2$. Prior to infiltration, 150 µM acetosyringone was added to the bacterial suspension. A blunt syringe was used to infiltrate the suspension into the tobacco leaves. Five days after infiltration, the leaves were harvested. For collection of apoplastic fractions, the harvested leaf material was washed three times with cold double-distilled H_2O, collected in ice-cold infiltration buffer (300 mM NaCl, 50 mM sodium phosphate buffer, pH 7.0) and vacuum-infiltrated twice at 7 kPa for 1 min. Subsequently, the leaves were blotted dry and transferred to a syringe barrel filled with glass wool and centrifuged for 10 min at 4°C and 950 g. The resulting protein solution was centrifuged at 16 000 g for an additional 10 minutes. His-tagged SBT3.8 was purified from the supernatant on Ni-NTA agarose as described above. SBT3.8 was bound to the column for 2 h, the column was washed extensively in buffer containing 20 mM imidazole, and SBT3.8 was eluted in buffer containing 400 mM imidazole. The eluate was dialyzed three times against 300 mM NaCl, 50 mM sodium phosphate buffer, pH 5.5, and used for further analysis.

Cleavage of peptide precursors by SBT3.8 and identification of cleavage products

Approximately 10 µg of recombinant propeptide and 100 ng of SBT3.8, or the same volume of the mock-purified fraction from empty-vector-infiltrated plants as control, were incubated at room temperature for 1 min up to 20 hours as indicated in the figures, or, if mass spectrometry (MS) analysis followed, for 30 min, in a total volume of 30 µl of 50 mM sodium phosphate buffer, pH 7.0 or 5.5, supplemented with 10 mM NaCl. To confirm the involvement of SBTs in PSK1 precursor cleavage, cleavage assays were performed with wild-type exudates pre-incubated for 30 min with 90 ng recombinant EPI10, which was expressed and purified as described previously with minor modifications (Schardon et al., 2016). For subsequent MS analysis, reactions were stopped by incubation at 95 °C for 10 min. If indicated in the figure legends, peptides were separated by ultrafiltration with a molecular mass cutoff of 10 kDa. Peptides were purified on C18 reverse-phase stage tips (Agilent, Santa Clara, CA, USA). C18 discs were conditioned in 80% (v/v) acetonitrile (ACN) and equilibrated in 0.5% (v/v) acetic acid by centrifugation at 5000 rpm for 1 min each. Peptides were bound by centrifugation at 3000 rpm for 1 min, washed in 0.5% (v/v) acetic acid for 1 min, and eluted with 50% (v/v) ACN. The isolated peptides were dried by vacuum evaporation. For subsequent SDS-PAGE analysis, the reaction was stopped by incubation at 95 °C for 10 min including loading dye. Samples were separated by SDS-PAGE or Tris-Tricine-PAGE (Schägger, 2006). Wild-type and *sbt3.8* mutant exudates were obtained by growing seedlings in submerged culture in 0.5 ×Murashige and Skoog medium with 1% (w/v) sucrose for 10 days as described (Ohyama et al., 2008). Under these conditions, seedlings release their extracellular protein content into the medium (Ohyama et al., 2008). Apoplastic proteins were enriched by ultrafiltration with a molecular mass cutoff of 5 kDa. Exudates were used at a protein equivalent of 500 ng for cleavage assays. For quantification, gels of the cleavage assays were scanned and the band area was determined by using ImageJ. Results are given as the percentage of the control for each lane of the gels.

Immunodetection

For western blots, proteins were transferred to nitrocellulose membranes using a semi-dry blotter and standard protocols (BioRad). GFP-fusion proteins were detected by polyclonal α-GFP antiserum (Thermo Fisher Scientific, Waltham, MA, USA). Horseradish peroxidase-conjugated anti-rabbit IgG was used as secondary antibody, followed by enhanced chemiluminescence detection using a Li-Cor Odyssee imager (Li-Cor Biosciences, Bad Homburg, Germany).

Mass spectrometry

Peptides were analyzed via LC-MS/MS using the nanoflow EASY-nLC 1000 System (Thermo Fisher Scientific) for HPLC separation and an Orbitrap hybrid mass spectrometer (Q-Exactive, Thermo Scientific, Waltham, MA, USA) for mass analysis. Peptides were eluted from a 75 µm analytical C18 column (PepMan, Thermo Fisher Scientific) on a linear gradient from 4% (v/v) to 64% (v/v) ACN over 135 min, and sprayed directly into the Q-Exactive Plus mass spectrometer. Proteins were identified via MS/MS on the basis of the fragmentation spectra of multiple charged peptides. Up to 12 data-dependent MS/MS spectra were acquired in the

Orbitrap for each full-scan spectrum. Full-scan spectra were acquired at 70 000 full-width half-maximum resolution, and fragment spectra at a resolution of 35 000 full-width half-maximum. An inclusion list was used for fragmentation containing peptides covering the C-terminus of the precursor, which includes the mature PSK pentapeptide. For tryptic in-gel digest, bands were excised and digested with proteomics-grade trypsin (Roche, Penzberg, Germany) according to (Shevchenko et al., 1996), and subjected to nano-LC-MS/MS analysis.

Real-time PCR

RNA isolation was performed as previously described with minor modifications (Kutschmar et al., 2009). cDNA was synthesized using up to 5 µg of total RNA, oligo dT primers, and RevertAid Reverse Transcriptase (Thermo Fisher Scientific). SBT primers for quantitative real-time PCR analysis were described and used previously; they are listed in Supplementary Table S2 (Stührwohldt et al., 2020b). Quantitative PCRs (qPCRs) (total volume 25 µl) were performed in biological triplicates with two technical repeats on the obtained cDNAs using a CFX96 Real-Time PCR Detection system (BioRad). Primer efficiencies and optimal primer concentrations were determined experimentally. qPCR was performed with *Taq* polymerase expressed in and purified from *E. coli* and SYBR Green (Cambrex Bio Science Rockland Inc., Rockland, ME, USA). Relative mRNA levels were determined after normalization to three reference genes (*Actin2*, *EF*, and *Tubulin*) following the equation described by Pfaffl (2001).

Results

To address the potential involvement of SBTs and PSK signaling in drought stress responses in Arabidopsis, the regulation of *SBT* and *PSK* gene expression in response to osmotic stress was analyzed. Seedlings were grown for 5 days on control medium and then transferred to mannitol-containing medium. Gene expression was compared in seedlings exposed to osmotic stress and control seedlings by qPCR. We observed that four out of five *PSK* genes (*PSK1*, *PSK3*, *PSK4*, and *PSK5*) and three out of 56 *SBT* genes (*SBT1.4*, *SBT3.7*, and *SBT3.8*) were significantly up-regulated in response to mannitol treatment (Fig. 1B). Induction levels for these *SBT* genes were similar to those for genes previously shown to be regulated under osmotic stress conditions (Sham et al., 2015), including *At3g14067* (which is the *SBT1.4* gene), *At3g46280*, and *At2g42540* (Fig. 1B). Minor differences in expression were detected for some additional *SBT* genes (Supplementary Fig. S1) but not investigated further.

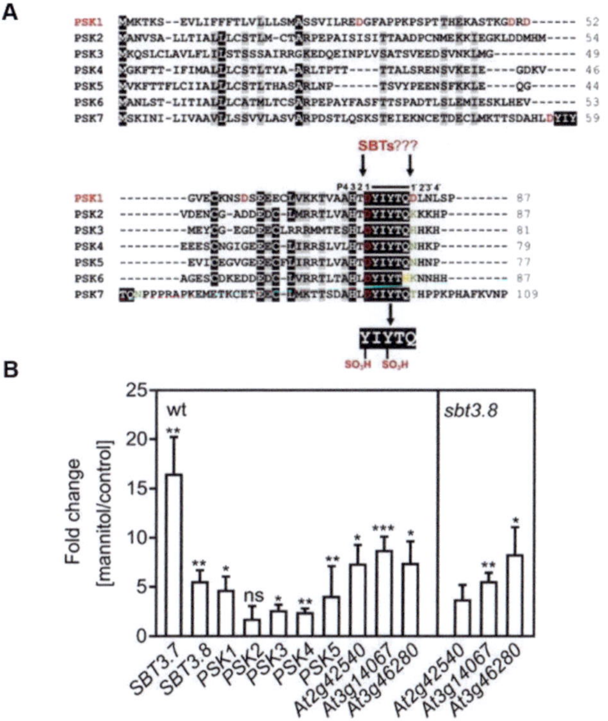

Figure 1. *SBT1.4*, *SBT3.7*, *SBT3.8*, and PSK precursor genes are up-regulated upon osmotic stress treatment. (A) Alignment of all seven Arabidopsis PSK precursor proteins. Entire protein sequences were aligned with ClustalW software. Identical amino acids conserved in at least six out of the seven sequences are shaded in black, and similar amino acids are shaded in grey. Conserved aspartate (Asp, D) residues flanking the mature PSK sequence are in red text, whereas alternative flanking amino acids are in green. Additional Asp residues in the PSK1 precursor are also marked in red. P1 to P4 and P1′ to P4′ indicate amino acid positions upstream and downstream of the cleavage sites, respectively. (B) Relative expression of *At2g42540*, *At3g14067* (alias *SBT1.4*), *At3g46280*, *SBT3.7*, *SBT3.8*, *PSK1*, *PSK2*, *PSK3*, *PSK4*, and *PSK5* in wild-type and *sbt3.8* loss-of-function mutant plants upon mannitol treatment. Plants were grown for 5 days under control conditions and transferred to plates containing no or 300 mM mannitol for 6 additional days. qPCR was performed on three biological replicates with two technical repeats, and gene expression was normalized to three reference genes and is shown as fold change in stressed versus

control plants. Each time point included pooled plant material of several independent plants. Significant differences compared with the control are indicated with asterisks (*P<0.05, **P<0.01, ***P<0.001; two-tailed *t*-test).

Next, we analyzed the osmotic stress tolerance of *sbt1.4*, *sbt3.7*, and *sbt3.8* loss-of-function mutants. While the *sbt1.4* and *sbt3.7* mutants did not seem to be affected (Supplementary Fig. S2), *sbt3.8* was clearly more sensitive to mannitol treatment than the wild type (Fig. 2A). The difference in shoot fresh weight between wild-type and *sbt3.8* mutants was more pronounced after mannitol treatment (Fig. 2B), and root fresh weight was significantly reduced in the *sbt3.8* mutant but not in the wild type (Fig. 2C). When seedlings were transferred from control media to mannitol-containing media at 5 days of age, they did not have any lateral roots. The number of lateral roots produced during the following 6 days on mannitol-containing medium was significantly lower for the *sbt3.8* loss-of-function mutant compared with the wild type (Fig. 2D). We also observed that roots of the *sbt3.8* mutant were significantly shorter than wild-type roots under stress conditions, whereas no difference was observed under control conditions (Fig. 2E). The impaired root growth and lateral root development of the *sbt3.8* mutant on mannitol-containing media, and the more severe reduction of shoot and root biomass, suggest a role for SBT3.8 in the acclimation to osmotic stress. Despite the apparent reduction in stress tolerance, osmotic stress markers were induced in the *sbt3.8* mutant to levels comparable to the wild type (Fig. 1B). Therefore, SBT3.8 does not seem to be required for osmotic stress perception, but may rather be involved in downstream signaling events. As PSK is known to promote root growth and lateral root development (Kutschmar et al., 2009), and since four out of seven *PSK* genes were up-regulated upon drought/osmotic stress treatment, we applied the disulfated, mature PSK pentapeptide in a seedling growth bioassay to assess the potential contribution of PSK to stress tolerance. Again, *sbt3.8* mutants were more sensitive to mannitol treatment, showing a stronger reduction in root and shoot growth than the wild type (Fig. 3A). The growth of both wild-type and *sbt3.8* seedlings was improved upon the addition of PSK (Fig. 3A) consistent with a role for PSK in stress mitigation. Again, we quantified the number of lateral roots and the primary root length after transfer to media containing mannitol (Fig. 3B, C). The number of lateral roots was increased by PSK treatment in both wild type and *sbt3.8*. However, the PSK-induced increase in lateral root number was much greater in *sbt3.8* (+59%) compared to the wild type

(+15%) (Fig. 3B). Likewise, the effect of PSK treatment on root growth was more pronounced in the *sbt3.8* mutant (+33%) than in the wild type (+14%) (Fig. 3C), suggesting that endogenous levels of PSK may be reduced in *sbt3.8*. The data are thus consistent with impaired PSK formation in the *sbt3.8* mutant.

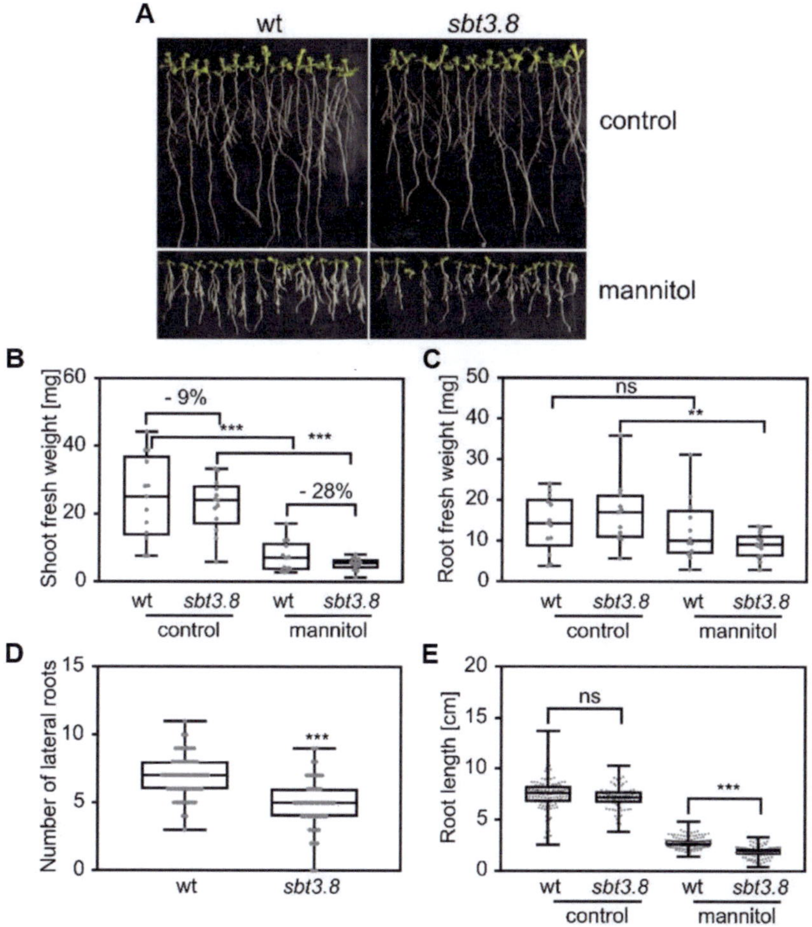

Figure 2. *sbt3.8* loss-of-function plants are more sensitive to osmotic stress. (A) Representative wild-type (wt) and *sbt3.8* plants grown under control conditions or with 300 mM mannitol. (B, C) Shoot (B) and root (C) fresh weight of wt and *sbt3.8* plants grown for 5 days under control conditions and transferred to plates containing no or 300 mM of mannitol for 6 additional days. (D) Number of lateral roots of wt and *sbt3.8* plants. (E) Root length of wt and *sbt3.8* plants under control conditions and after mannitol treatment. Data

are shown as the mean ±SE for one representative of at least three independent experiments: (B) n≥13, (C) n≥11, (D) n≥47, (E) n≥79). Asterisks show significant differences between conditions or genotypes as indicated (**$P<0.01$, ***$P<0.001$; two-tailed *t*-test).

PSK formation requires at least two steps of proteolytic processing to release the bioactive pentapeptide from its precursor (Fig. 1A). The *PSK* gene most strongly up-regulated in response to osmotic stress was *PSK1* (Fig. 1B). In contrast to other PSK precursors, in which the PSK pentapeptide is flanked by Asp only at the N-terminus, proPSK1 features Asp at both the N-terminal and C-terminal processing sites (Fig. 1A). Asp is located immediately upstream of the scissile bond (in position P1) at the N-terminal site, and downstream of the scissile bond (in position P1′) at the C-terminal site (Fig. 1A). Interestingly, SBT3.8 has been described as an Asp-specific protease (a phytaspase; (Chichkova et al., 2018)). SBT3.8 was found to be highly selective for Asp at acidic pH (pH 5.5), cleaving a panel of synthetic peptide substrates with Asp in P1 (Chichkova et al., 2018), and the CLEL6 precursor with Asp in P1′ (Stührwohldt et al., 2020b). We thus tested whether proPSK1 is a substrate of SBT3.8 and, if so, which site is cleaved by SBT3.8.

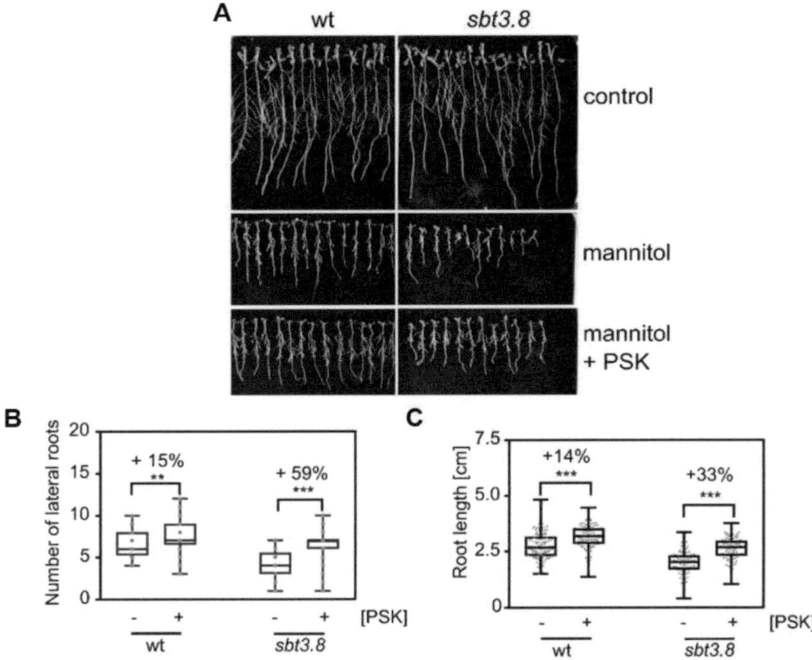

Figure 3. PSK can recover the mannitol-induced *sbt3.8* loss-of-function phenotype. (A) Representative pictures of wild-type (wt) and *sbt3.8* plants grown under control conditions, with 300 mM mannitol, or with 300 mM mannitol and 1 µM PSK. (B, C) Number of lateral roots (B) and root length (C) of wt and *sbt3.8* plants grown as in (A). Data are shown as the mean ±SE for one representative experiment: (B) n≥49, (C) n≥74. Asterisks indicate significant differences (**P<0.01, ***P<0.001; two-tailed t-test).

SBT3.8 was expressed with a C-terminal hexa-His tag in tobacco (*N. benthamiana*) leaves and purified from apoplastic wash fluids by affinity chromatography as described (Stührwohldt et al., 2020b). C-terminally hexa-His-tagged proPSK1 was expressed in *E. coli* and purified by affinity chromatography, ultrafiltration, and size exclusion chromatography (Fig. 4A). Cleavage of recombinant proPSK1 was detected by SDS-PAGE after incubation with SBT3.8. Degradation was much reduced in a control digest, in which SBT3.8 was replaced by mock-purified tobacco apoplastic fluid (Fig. 4B). To test whether SBT3.8 is required for proPSK1 processing in planta, we incubated recombinant proPSK1 with exudates of wild-type and *sbt3.8* mutant seedlings grown in submerged culture. Rapid (within 5 min) degradation was observed for wild-type but not *sbt3.8* exudates (Fig. 4C), indicating that cleavage requires SBT3.8, and that there are no redundant activities, at least not at this stage of seedling development. In order to verify that the observed cleavage was indeed caused by SBTs, we repeated the experiment with wild-type exudates pre-treated for 30 minutes with recombinant Extracellular Protease Inhibitor10 (EPI10), a specific inhibitor of subtilases (Tian et al., 2005). EPI10 almost completely inhibited cleavage, indicating that SBTs are responsible for the observed degradation of proPSK1 (Fig. 4D).

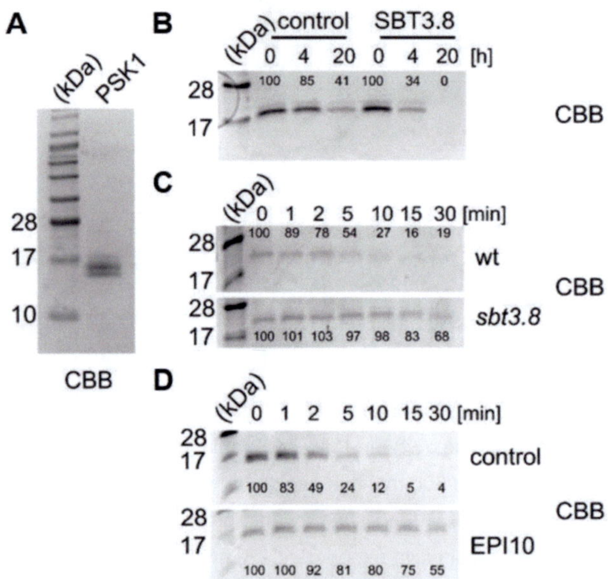

E

	pH 7	pH 5.5
	...CLVKKTVAAHTDYIYTQDLNLSPHHHHHH	CLVKKTVAAHTDYIYTQDLNLSPHHHHHH
control	no peptides detectable	no peptides detectable
SBT3.8	TVAAHTDYIYTQ (61.74)	HTDYIYTQ (35.44)
	VAAHTDYIYTQ (39.78)	AAHTDYIYTQ (41.59)
	AAHTDYIYTQ (55.96)	AHTDYIYTQ (43.58)
	...CLVKKTVAAHTDYIYTQALNLSPHHHHHH	CLVKKTVAAHTDYIYTQALNLSPHHHHHH
control	no peptides detectable	no peptides detectable
SBT3.8	no peptides detectable	no peptides detectable

Figure 4. C-terminal processing of the PSK precursor by SBT3.8 depends on Asp in P1′. (A) PSK1 precursor expressed in *E. coli* and purified by Ni-NTA chromatography,

ultrafiltration, and size exclusion chromatography. Proteins were separated by Tris-Tricine-PAGE and stained with Coomassie Brilliant Blue (CBB). (B) proPSK1 is processed by SBT3.8 *in vitro*. The recombinant PSK1 precursor was incubated with purified SBT3.8 or a mock-purified control fraction (100 ng for SBT3.8 and equivalent volumes for the mock-purified control) for the times indicated. Proteins were separated on Tris-Tricine gels and stained with CBB. (C) proPSK1 processing depends on SBT3.8. Recombinant proPSK1 was treated with exudates of wild-type (wt) or *sbt3.8* loss-of-function plants (500 ng protein equivalent of exudates for each treatment) for the times indicated, separated on Tris-Tricine gels, and stained with CBB. (D) Cleavage activity is inhibited by EPI10. Cleavage assays with recombinant proPSK1 were performed with wt seedling exudate, and with exudate pre-treated with EPI10 for 30 min (500 ng protein aliquots of exudates for each treatment). Reactions were stopped at the time indicated, separated on Tris-Tricine gels, and stained with CBB. Cleavage assays were repeated at least three times with similar results. Quantification of the cleavage products is shown for each lane as the percentage of the control band area. (E) SBT3.8 cleaves proPSK1 in an Asp-dependent manner at the C-terminal site irrespective of pH. Peptides released by SBT3.8 or the mock-purified control from recombinant proPSK1 at pH 7 or 5.5 were identified by mass spectrometry (top). No peptides were identified from the region of interest when Asp in P1′ was substituted by Ala (bottom). Spectra of identified peptides are shown in Supplementary Fig. S4.

The absence of distinct processing products in *in vitro* digests of proPSK1 with SBT3.8 (Fig. 4B, C) may be explained by the presence of multiple Asp residues upstream of the PSK sequence (marked in red in Fig. 1A) resulting in SBT3.8-mediated degradation of the precursor. Therefore, to confirm precursor processing at either one of the two Asp residues flanking the PSK pentapeptide, we replaced the four Asp residues in the pro-part of the precursor with alanine (Ala) residues (Supplementary Fig. S3A). The Ala-substituted precursor was expressed in *E. coli* and subjected to cleavage by SBT3.8 (Supplementary Fig. S3B). Distinct cleavage products were now detected after SDS-PAGE, cut out from the gel, digested with trypsin, and analyzed by MS. Identification of the semi-tryptic peptide K/TVAAHTDYIYTQ/D (cleavage sites shown by '/', PSK sequence in italic text) indicated that the precursor was cleaved by SBT3.8 at the right border of the PSK pentapeptide, releasing the C-terminal extension (Supplementary Fig. S3C).

In order to confirm the specificity of SBT3.8-mediated proPSK1 cleavage at the C-terminal but not the N-terminal processing site, the recombinant wild-type precursor was digested with SBT3.8 or the mock-purified negative control, the peptide fraction was recovered by ultrafiltration, and peptides were purified on C18 stage tips and analyzed by MS. We identified a series of peptides, all indicative of cleavage at the C-terminal processing site (Fig. 4E; peptide spectra are shown in Supplementary Fig. S4). N-termini were somewhat variable and likely resulted from non-specific cleavage by contaminating tobacco protease(s) that are abundant in apoplastic wash fluids (Grosse-Holz et al., 2018). These peptides were detected specifically only in the SBT3.8 digest, irrespective of pH (7 or 5.5), and not for the control reaction (Fig. 4E; peptide spectra are shown in Supplementary Fig. S4). To further test the significance of Asp for C-terminal cleavage by SBT3.8, the cleavage assay was repeated under the same conditions with a site-directed proPSK1 mutant in which this Asp residue was replaced by Ala. MS failed to identify any peptides derived from C-terminal cleavage either at pH 7 or at pH 5.5 (Fig. 4E). These data indicate that SBT3.8 cleaves the PSK1 precursor specifically at the C-terminal processing site and that this cleavage depends on the presence of Asp in P1′. To test whether SBT3.8 is able to cleave other Asp-containing propeptides, we purified recombinant proRGF1 [the propeptide of Root Meristem Growth Factor 1 (RGF1), alias GOLVEN11/CLEL8] from *E. coli* and incubated it with exudates of wild-type or *sbt3.8* mutant seedlings (Supplementary Fig. S5A, B). RGF1 is a tyrosine-sulfated peptide that also features Asp in P1′ of the processing site and is required for the maintenance of the root stem cell niche (Matsuzaki et al., 2010). proRGF1 was cleaved by wild-type but not by *sbt3.8* exudates, indicating that SBT3.8 is required for the observed processing of proRGF1. A site-directed D104A mutant of proRGF1 was resistant to cleavage by SBT3.8 (Supplementary Fig. S5B), thus confirming cleavage at D104 and the Asp specificity of processing.

To try to verify that Asp-specific cleavage is required for PSK formation *in vivo*, we generated transgenic Arabidopsis plants overexpressing wild-type proPSK1 from rice in translational fusion with GFP (proPSK1-GFP), and a site-directed mutant in which the two flanking Asp residues had been replaced by Ala (proPSK1-DD/AA). The functionality of rice proPSK1-GFP in Arabidopsis was confirmed in bioassays based on the growth-promoting activity of PSK (Kutschmar et al., 2009; Stührwohldt et al., 2011). Root length was increased by more than 50% and hypocotyls were significantly longer in three independent transgenic lines expressing proPSK1-GFP compared with Col-0 wild-type plants (Fig. 5A, B). If Asp-specific cleavage were required for PSK biogenesis, we would expect the Ala-substituted proPSK1 mutant to be inactive in this assay. However, the root and hypocotyl length of three independent *proPSK1-DD/AA-GFP* lines was increased to a similar extent as that in plants expressing the proPSK1-GFP wild-type construct (Fig. 5A, B), even though processing was impaired by the D-to-A cleavage site mutation (Fig. 5C). To reconcile the requirement for Asp at the C-terminal processing site for PSK1 formation (Figs 4E, 5C) with the apparent activity of the D-to-A substituted proPSK1 mutant *in vivo* (Fig. 5A, B), we compared the expression of endogenous *PSK* genes in the transgenic plants and in wild-type plants. Interestingly, *PSK1*, *PSK2*, and *PSK3* were found to be up-regulated in plants carrying the rice proPSK1 overexpression (ox) constructs (Fig. 5D), with particularly strong induction of PSK1. The observed increase in root and hypocotyl length in plants expressing non-cleavable proPSK1-DD/AA may thus be caused by the up-regulation of endogenous PSKs. A possible explanation for the up-regulation of endogenous PSKs is the apparent cleavage of the proPSK1-DD/AA mutant by SBT3.8 (or other proteases) at different sites (including the Asp residues that are cleaved by SBT3.8 upstream of the PSK sequence; Supplementary Fig. S3) producing longer versions of the PSK peptide that may retain (some of) the activity of the mature PSK pentapeptide. This residual activity may be sufficient to induce the expression of endogenous *PSK1*, *PSK2*, and *PSK3* genes, all resulting in the formation of correctly processed and bioactive PSK.

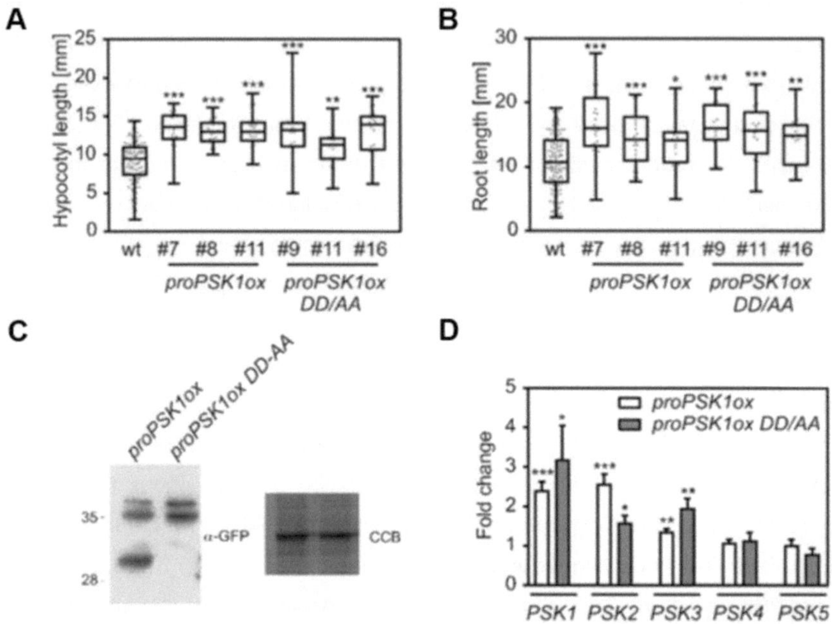

Figure 5. Root and hypocotyl growth are induced in plants expressing proPSK or the Asp-substituted proPSK mutant. (A, B) Hypocotyl (A) and root (B) length of three independent lines overexpressing proPSK1-GFP or PSK1-DD/AA-GFP in comparison to the wild type (wt). Experiments were performed three times with similar results. Data are shown as the mean ±SE for one representative experiment: (A) n≥16, (B) n≥18). Significant differences from the wt are indicated with asterisks (*P<0.05, **P<0.01, ***P<0.001; two-tailed *t*-test). (C) The DD/AA mutation impairs proPSK1 processing *in vivo*. Leaf protein extracts (20 µg) of representative PSK1ox-GFP or PSK1ox DD/AA-GFP lines were analyzed by anti-GFP immunoblotting. A duplicate Coomassie Brilliant Blue-stained gel is shown as a loading control (right). (D) qPCR analysis of plants overexpressing PSK1 or PSK1 DD/AA precursors. qPCR was performed on RNA isolated from 11-day-old plants in three biological replicates with two technical repeats. Gene expression was normalized to three reference genes and is shown as fold change in plants overexpressing PSK1 or PSK1 DD/AA precursors versus wt plants. Each time point included pooled plant material of several independent plants. Significant differences from the wt are indicated with asterisks (*P<0.05, **P<0.01, ***P<0.001; two-tailed *t*-test).

We further used the *proPSK1ox* lines to directly assess the effect of PSK on osmotic stress tolerance. Experimental plants were grown for 5 days on control media, then transferred to either control medium or mannitol-containing medium, and 6 days later, the increment of root growth was measured (Fig. 6A). The root growth of wild-type plants was severely impaired on mannitol-containing medium (Fig. 6B), whereas *proPSK1ox* plants coped better with osmotic stress. The increment of root growth of *proPSK1ox* plants was about twice as large as that in wild-type plants (Fig. 6B). Further, we quantified the number of lateral roots, and recorded significantly higher numbers of lateral roots in *proPSK1ox* lines (Fig. 6C). These data confirm a role for PSK in stress mitigation and, consistent with the up-regulation of *PSK1*, *PSK2*, *PSK3*, and *PSK4* genes under stress (Fig. 1), they indicate that the level of stress tolerance is limited by *PSK* expression levels.

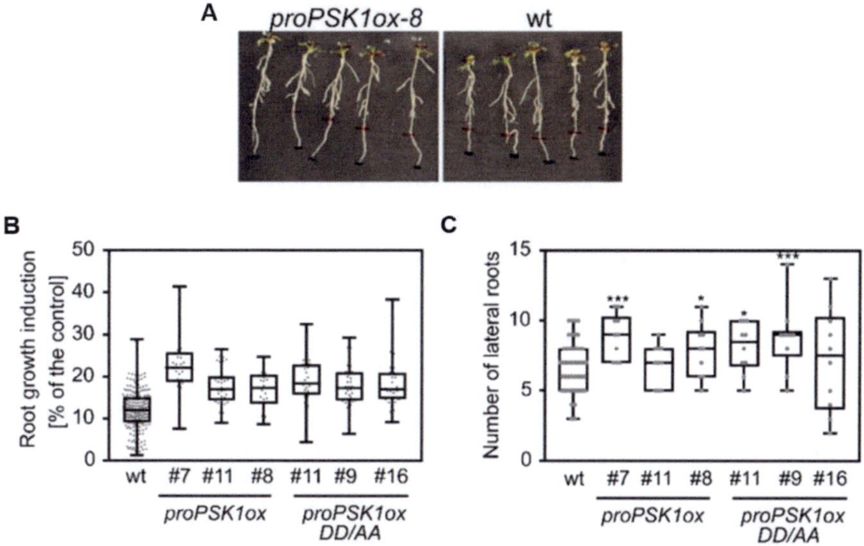

Figure 6. Overexpression of *PSK1* improves osmotic stress tolerance. (A) Representative *proPSK1ox-8* and wild-type (wt) plants 6 days after transfer to mannitol-containing medium. The increment of root growth is marked. (B) Root growth increment of wt and PSK overexpressing plants after transfer to mannitol-containing medium, shown as a percentage of the root growth increment of plants grown on control medium. (C) Number of lateral roots formed by overexpressors of proPSK1 or proPSK1-DD/AA compared with the wt. Data are shown as the mean ±SE for one representative of three independent experiments: (B) n≥10, (C) n≥10). Significant differences from the wt are indicated with asterisks (*P<0.05, ***P<0.001; two-tailed *t*-test).

The findings that *SBT* expression is induced upon osmotic stress (Fig. 1), stress sensitivity is increased in the *sbt3.8* mutant (Fig. 2B, C), and SBT3.8 is able to process the PSK1 precursor (Fig. 4) suggest that osmotic stress acclimation is regulated not only at the level of *PSK* expression, but also at the post-translational level of proPSK processing. To corroborate this notion and to confirm SBT3.8 as a limiting factor in stress mitigation, we generated transgenic Arabidopsis plants overexpressing SBT3.8 as a C-terminal sfGFP fusion. Overexpression of SBT3.8 was confirmed by qPCR and immunoblot analysis (Supplementary Fig. S6A, B). Under control growth conditions, 5-day-old SBT3.8ox seedlings had significantly longer main roots than wild-type seedlings (Fig. 7A), comparable to seedlings overexpressing proPSK1 (Fig. 5B). On mannitol-containing medium, SBT3.8ox plants showed greater osmotic stress tolerance than wild-type plants (Fig. 7B–F). The number of lateral roots (Fig. 7B) and shoot and root fresh weight (Fig. 7D, E) were all increased in SBT3.8ox plants compared to the wild type. SBT3.8ox plants also performed better than the wild type when grown on soil under continuous drought stress (Supplementary Fig. S6C). Finally, we analyzed the expression of genes involved in abscisic acid (ABA) biosynthesis, including *ABA1* (zeaxanthin epoxidase), *ABA2* (cytosolic short-chain dehydrogenase/reductase forming ABA-aldehyde from xanthoxin), and AAO3 (aldehyde oxidase delta isoform catalyzing the final step in ABA biosynthesis). Under control conditions, the expression of *ABA1*, *ABA2*, and *AAO3* was unchanged in SBT3.8 overexpressors compared with the wild type (Fig. 7F). Upon osmotic stress treatment, *ABA1*, *ABA2*, and *AAO3* expression was induced to higher levels in *SBT3.8* overexpressors compared with the wild type, although not all differences were statistically significant (Fig. 7G). Increased ABA production may explain why *SBT3.8* overexpressors cope better with osmotic stress. The data support a regulatory role for SBT3.8 in the production of PSK as a signal for osmotic and drought stress acclimation.

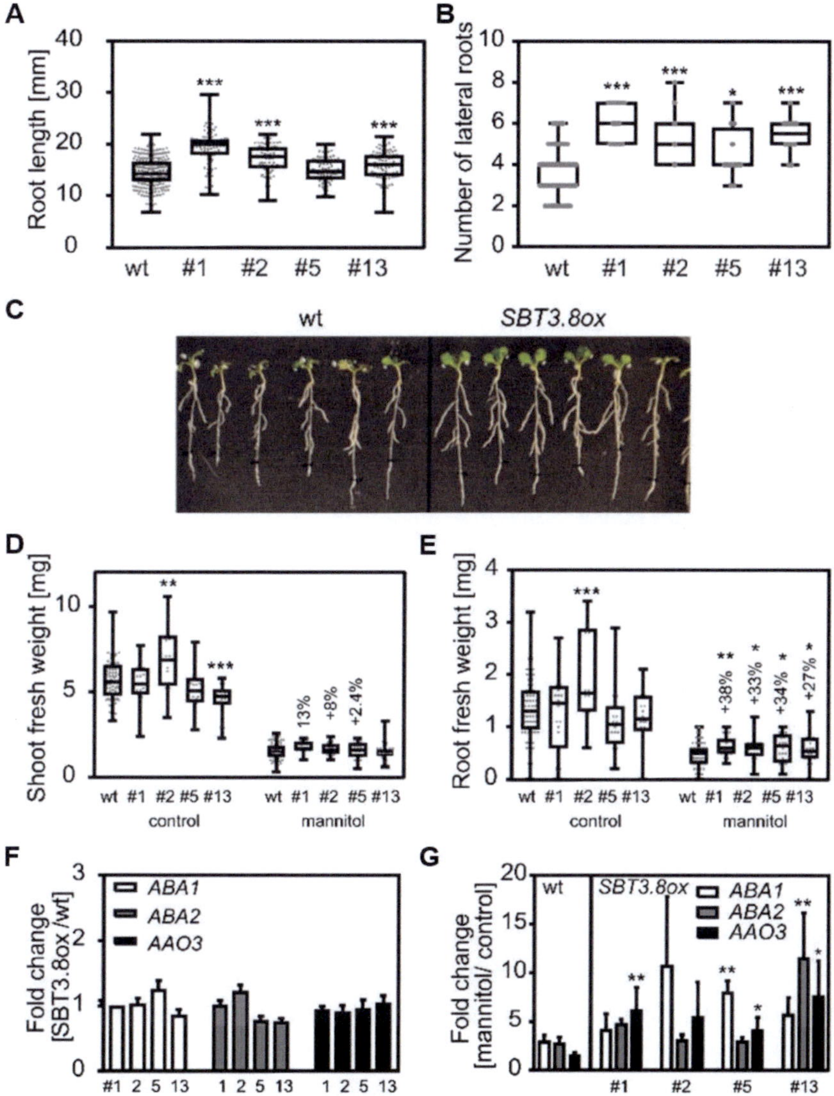

Figure 7. Constitutive overexpression of *SBT3.8* improves osmotic stress tolerance and induces growth. (A) Root length is significantly increased in four independent *SBT3.8*-overexpressing (*SBT3.8ox*) seedlings compared with the wild type (wt). (B) The number of lateral roots is increased in four independent *SBT3.8ox* lines compared with the wt. (C) Representative images of wt and *SBT3.8ox* plants subjected to mannitol stress. (D, E) Shoot (D) and root (E) fresh weight of plants grown on mannitol-containing medium are

greater in *SBT3.8ox* plants compared with wt plants. Experiments were performed three times with similar results. Data are shown as the mean ± SE for one representative experiment: (A) n≥66, (B) n≥12, (C) n≥18, (D) n≥17). Significant differences from the wt are indicated with asterisks (*P<0.05, **P<0.01, ***P<0.001; two-tailed *t*-test). (F, G) Relative expression of *ABA1*, *ABA2*, and *AAO3* in wt and *SBT3.8ox* lines under control (F) and osmotic stress (G) conditions. Relative expression is shown as fold change in *SBT3.8ox* versus wild-type plants (F) or stressed versus control plants (G). Plants were grown for 5 days under control conditions and transferred to plates containing no or 300 mM mannitol for 6 additional days. qPCR was performed on pooled plant material in three biological replicates with two technical repeats. Gene expression was normalized to three reference genes. Significant differences from the wt are indicated with asterisks (*P<0.05, **P<0.01; two-tailed *t*-test).

Discussion

PSK as a drought stress signal in Arabidopsis

PSK is known primarily for its growth-regulatory properties. It was identified in 1996 as a mitogenic growth factor that stimulates cell division in asparagus cell cultures (Matsubayashi and Sakagami, 1996). In Arabidopsis, PSK induces root and hypocotyl growth. The elongation of roots and hypocotyls is induced to similar levels by PSK treatment, by overexpression of the PSK1 precursor, and by overexpression of the PSK receptor PSKR1 (Kutschmar et al., 2009; Stührwohldt et al., 2011; Hartmann et al., 2013). These growth responses were found to rely on cell expansion rather than cell division (Matsubayashi et al., 2006; Kutschmar et al., 2009; Stührwohldt et al., 2011; Ladwig et al., 2015). However, in addition to its growth-regulatory properties, PSK is also involved in pollen germination and pollen tube elongation (Chen et al., 2000; Stührwohldt et al., 2015), vascular development (Holzwart et al., 2018), and the regulation of plant immunity (Loivamäki et al., 2010; Igarashi et al., 2012; Mosher et al., 2013; Rodiuc et al., 2016; Zhang et al., 2018). Most recently, PSK was found to act as a peptide signal for stress-induced flower drop in tomato (Reichardt et al., 2020). Upon drought stress, PSK precursor genes and the precursor-processing subtilase SlPhyt2 are up-regulated in the pedicel proximal to the abscission zone. Mature PSK acts in the abscission zone, where it induces the expression of cell-wall hydrolases to initiate the cell separation process (Reichardt et al., 2020). These findings prompted us to investigate whether PSK precursors and cognate subtilases may also be involved in drought stress responses in Arabidopsis.

We found that three subtilase genes (*SBT1.4*, *SBT3.7*, and *SBT3.8*) and four PSK precursor genes (*PSK1*, *PSK3*, *PSK4*, and *PSK5*) are up-regulated in Arabidopsis seedlings in response to osmotic stress (Fig. 1B). Tolerance to drought stress and osmotic stress was compromised in *sbt3.8* mutants and the latter could be restored by PSK treatment. In contrast, osmotic stress tolerance was improved in transgenic plants overexpressing either SBT3.8 or proPSK1. These data imply that PSK signaling acts in drought stress mitigation in Arabidopsis, and suggest that SBT3.8 is a limiting factor in PSK formation.

PSK precursor processing

The only protease previously implicated in PSK processing in Arabidopsis is SBT1.1. This protease was shown to cleave a synthetic peptide derived from the PSK4 precursor *in vitro* (Srivastava et al., 2008). SBT1.1-dependent processing of proPSK4 was also observed *in planta*, at least when the precursor was expressed ectopically under the control of the constitutive CaMV 35S promoter. However, the SBT1.1 cleavage site is located several residues upstream of the PSK sequence and, therefore, cleavage by SBT1.1 does not produce the mature and bioactive peptide (Srivastava et al., 2008). In addition, no PSK-related phenotypes have been reported for the *sbt1.1* mutant (Srivastava et al., 2008). Therefore, the contribution of SBT1.1 to PSK maturation, if any, is questionable.

In tomato plants, PSK maturation depends on phytaspase-2 (SlPhyt2; Reichardt et al., 2020). Phytaspases were first identified in tobacco and rice (Chichkova et al., 2018). They constitute a subgroup of subtilases that cleave their substrates in an Asp-specific manner (hence the name: phytaspase is short for plant Asp-specific protease). For substrate recognition and cleavage by phytaspases, Asp is required in the P1 position (immediately upstream of the scissile bond) (Chichkova et al., 2018; Reichardt et al., 2020). Consistently, SlPhyt2 cleaves PSK precursors at the N-terminal processing site where the P1 Asp is conserved in all eight PSK precursors in tomato, and in the seven Arabidopsis PSK precursors as well (Kaufmann and Sauter, 2019; Reichardt et al., 2020).

In contrast to the 12 phytaspases tentatively identified in tomato (Reichardt et al., 2020), a single gene, *SBT3.8*, has been suggested as the phytaspase orthologue in Arabidopsis. SBT3.8 cleaves substrate peptides specifically after Asp in P1 (Chichkova et al., 2018). Therefore, cleavage of PSK precursors by SBT3.8 would be expected at the N-terminus of the PSK peptide sequence. However, we observed processing of proPSK1 at the C-terminal site, which features Asp in the P1′ position (immediately downstream of the cleavage site), rather than in P1 (see Fig. 1A for the amino acid sequence of proPSK1). This Asp residue (D60 of proPSK1) turned out to be required for cleavage by SBT3.8 (Fig. 4E). Cleavage of proPSK1 at the N-terminal site was not observed either at physiological pH of 5.5, or at pH 7.5 (Fig. 4E), at which SBT3.8 was reported to exhibit less stringent substrate specificity (Chichkova et al., 2018).

Cleavage of proPSK1 at D60 is fully consistent with our previous report on the biogenesis of CLEL6 and CLEL9. The respective precursors were found to be cleaved by SBT3.8 in an Asp-specific manner, and Asp was required in P1′ for processing (Stührwohldt et al., 2020b). Likewise, we show here that cleavage of RGF1 (alias GOLVEN11/CLEL8) by SBT3.8 depends on D104 (Supplementary Fig. S5). In RGFs, the critical P1′ Asp residue marks the N-terminus of the mature peptides, whereas in proPSK1 the critical Asp is located C-terminally to the PSK pentapeptide. We may thus anticipate a broader role for SBT3.8 in the biogenesis of peptide hormones carrying P1′ Asp at either the N-terminal or the C-terminal processing site. With respect to PSK biogenesis, SBT3.8 appears to be specific for the PSK1 precursor, as this is the only one featuring Asp in P1′. Hence, other unknown proteases must be involved in the C-terminal processing of Arabidopsis proPSK2 to proPSK7. SBT1.4 and SBT3.7 are obvious candidates, as their expression is up-regulated in response to osmotic stress as well (Fig. 1B).

The protease(s) responsible for N-terminal processing of Arabidopsis PSK precursors also remain(s) to be identified. Whether the N-terminal processing activity is also up-regulated by drought is unknown. If it is, we may expect to find the protease(s) among the SBTs up-regulated in response to osmotic stress (Fig. 1B, Supplementary Fig. S1A). Similar to PSK maturation in tomato (Reichardt et al., 2020), a *bona fide* Arabidopsis phytaspase may be involved in this processing step. Alternatively, the introduction of a negative charge in P1′ by tyrosine sulfation may allow recognition of the N-terminal site by SBT3.8. Tyrosine sulfation is known to be required for full peptide activity (Matsubayashi et al., 2006; Kutschmar et al., 2009). Mutants impaired in TPST activity show severe developmental defects that can be rescued by the addition of the mature disulfated PSK peptide (Matsubayashi et al., 2006; Kutschmar et al., 2009; Stührwohldt et al., 2011). The requirement of tyrosine sulfation for bioactivity was rationalized by structural studies showing that the two sulfate moieties contribute to the interaction of PSK with its receptor PSKR1, thus sensitizing the perception of PSK by PSKR1 (Wang et al., 2015). The hypothetical contribution of sulfotyrosine to cleavage site recognition and peptide biogenesis provides another explanation for the strong phenotype observed in TPST loss-of-function mutants.

Complex regulation of osmotic stress responses by peptides

In addition to PSK, other peptides are known to contribute to plant responses to osmotic stress. A cysteine-rich peptide in rice, OsDT11, enhances drought tolerance in an ABA-dependent manner (Li et al., 2017). Transgenic plants overexpressing OsDT11 show enhanced drought tolerance, reduced water loss, reduced stomatal density, and increased levels of ABA (Li et al., 2017). PROPEPs are primarily known as precursors of immune-regulatory phytocytokines (Huffaker and Ryan, 2007). However, one family member, Arabidopsis PROPEP3, was recently shown to be involved in salt stress responses (Nakaminami et al., 2018). The expression of AtPROPEP3 increases in response to salt stress, and salt tolerance correlates with AtPROPEP3 expression levels in transgenic plants. Stress tolerance is mediated by a fragment of the PEP3 phytocytokine and depends on the PEP receptor PEPR1 (Nakaminami et al., 2018). Likewise, CAP-derived peptide (CAPE), which is derived from the C-terminus of tomato pathogenesis-related protein 1 (PR1), was also first identified as an immune-modulatory peptide (Chen et al., 2014). In Arabidopsis, CAPE1 negatively regulates salt tolerance (Chien et al., 2015). Also in Arabidopsis, drought stress was shown to induce the production of CLE25 as a long-distance signal for increased resistance to dehydration. CLE25 binds to the BAM3 receptor in the leaves and regulates stomatal aperture in an ABA-dependent manner (Takahashi et al., 2018). Here we add PSK to the suite of drought stress signals in Arabidopsis, and SBT3.8 as a limiting factor for PSK production. Which proteases are involved, and the extent to which they contribute to the production of the other peptides, remains to be seen. Their number may well exceed the number of peptides they produce, thus highlighting the complexity of peptide-mediated osmotic stress responses.

Supplementary Figures

A

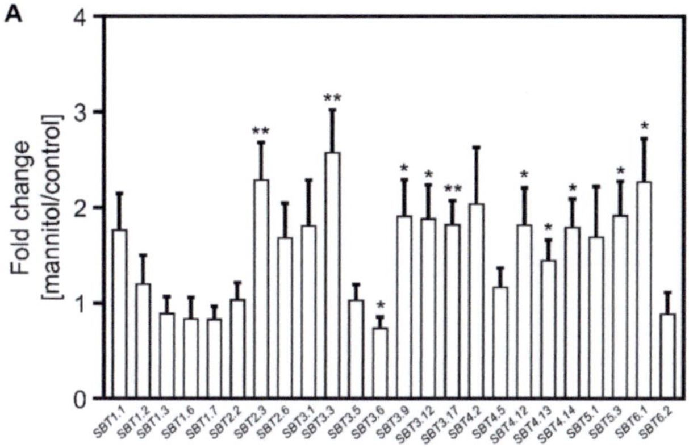

B

subtilase	expression level
SBT1.5	not detectable
SBT1.8	not detectable
SBT1.9	not detectable
SBT2.1	not detectable
SBT2.4	not detectable
SBT2.5	not detectable
SBT3.2	not detectable
SBT3.4	not detectable
SBT3.10	not detectable
SBT3.11	not detectable
SBT3.13	Not detectable after mannitol treatment
SBT3.14	not detectable
SBT3.15	not detectable
SBT3.16	not detectable
SBT3.18	not detectable
SBT4.1	not detectable
SBT4.4	not detectable
SBT4.5	not detectable
SBT4.7	not detectable
SBT4.8	not detectable
SBT4.9	not detectable
SBT4.11	not detectable
SBT4.13	not detectable
SBT4.15	not detectable
SBT5.2	not detectable
SBT5.4	not detectable
SBT5.5	not detectable
SBT5.6	not detectable

Supplementary Figure S1: Analysis of SBT gene expression after osmotic stress treatment. Stress responsive expression of all the SBT genes not shown in figure 1. (A) For SBTs that were detectable, fold induction is shown as the ratio of gene expression of stressed over control plants. (B) SBTs with expression levels below the detection limit in both conditions, or after mannitol treatment. Plants were grown for 5 days under control conditions and transferred to plates containing no or 300 mM of mannitol for six additional days. qPCR was performed on three biological replicates with two technical repeats, and gene expression was normalized to three reference genes and is shown as fold change in stressed vs control plants. Each time point included pooled plant material of several independent plants. * and ** indicate significant differences to the wild type at $P < 0.05$ or $P < 0.01$, respectively (two-tailed t test).

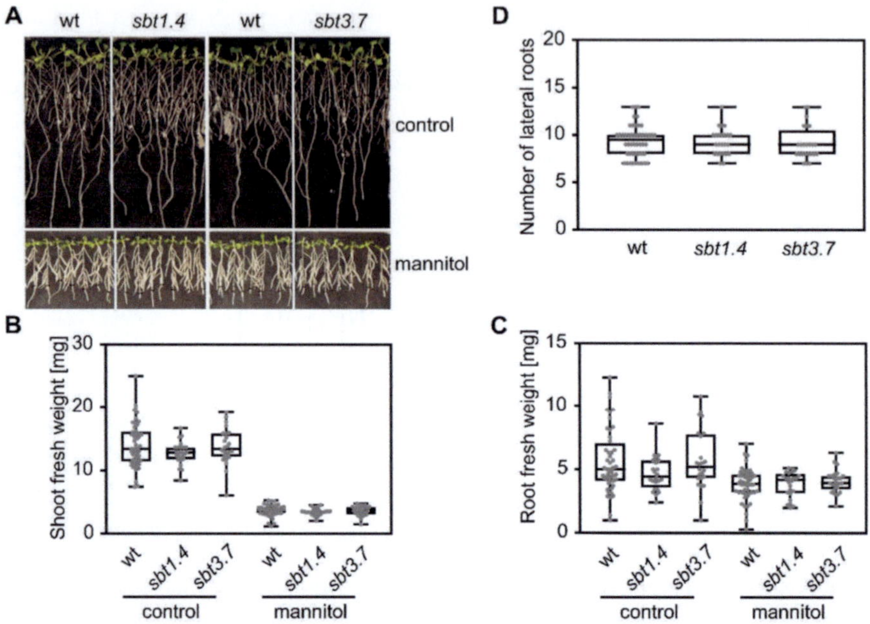

Supplementary Figure S2: sbt1.4 and sbt3.7 loss-of-function plants do not show changes in sensitivity to osmotic stress. (A) Representative wild-type (wt), sbt1.4 and sbt3.7 plants grown under control conditions or with 300 mM mannitol. (B-C) Shoot (B) and root (C) fresh weight [mg] of wt, sbt1.4 and sbt3.7 plants grown for 5 days under control conditions and transferred to plates containing no or 300 mM of mannitol for six additional days. (D) Number of lateral roots of wt, sbt1.4 and sbt3.7. Data are shown for one representative of at least three independent experiments as the mean ± SE ((B) n ≥ 24, (C) n ≥ 24, (D) n ≥ 24). No significant differences were detected (two-tailed t test).

A

MGVILREAGFAPPKPSPTTHEKASTKGARAGVECKNSASEEECLVKKTV

AAHTDYIYTQDLNLSP

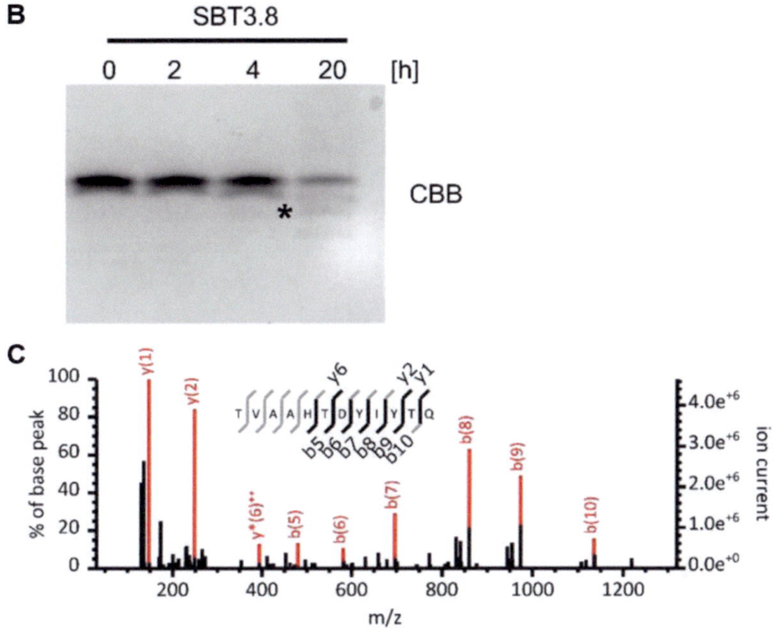

Supplementary Figure S3: Aspartate to alanine substitutions prevent PSK1 precursor cleavage. (A) Sequence of the PSK1 precursor that was created by aspartate to alanine substitutions at amino acid positions 8, 28, 30 and 38. Green letters indicate the substituted amino acids, red letters show aspartates flanking the PSK pentapeptide (YIYTQ, orange) in P1 at the N-terminal, and in P1' at the C-terminal cleavage site. (B) Mutated proPSK1 is processed by SBT3.8 *in vitro*. The recombinant PSK1 precursor was incubated with purified SBT3.8 (100 ng) for the time indicated. Proteins were separated on Tris-Tricine gels and stained by CBB. The black asterisk marks the cleavage product that was analyzed by mass spectrometry. (C) The protein cleavage product was digested with trypsin and the SBT3.8 cleavage site was identified at the aspartate in P1' at the C-terminal site as indicated by the spectrum of the peptide shown. Peptide identity is confirmed by the ions identified in the y- and b-series, respectively.

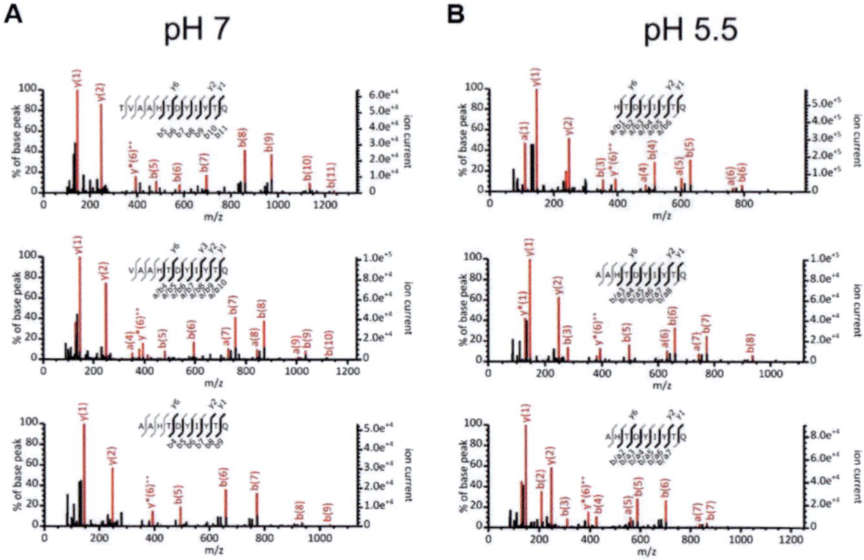

Supplementary Figure S4: Fragmentation spectra of peptides identified after PSK1 precursor cleavage by SBT3.8. Peptides released by SBT3.8 or the mock-purified control from recombinant proPSK1 at pH 7 or 5.5 were identified by mass spectrometry (Figure 4). Peptides were isolated by ultrafiltration (< 10 kDa MWCO) and solid-phase extraction, and analyzed by LC-ESI-MS/MS. The fragmentation spectra are shown for the peptides identified at (A) pH 7 and (B) pH 5.5. Peptide identities were confirmed by the ions identified in the y- and b-series, respectively.

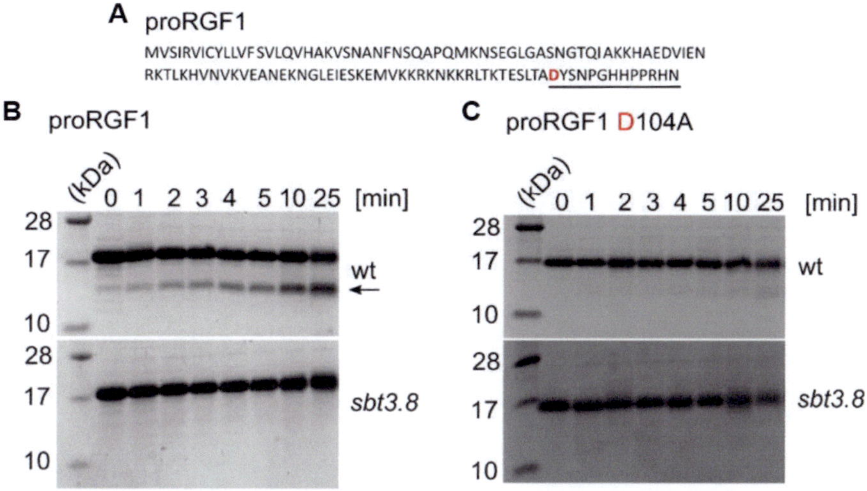

Supplementary Figure S5: RGF1 is cleaved by SBT3.8 in an Asp-dependent manner. (A) Amino acid sequence of proRGF1. The mature RGF1 peptide is underlined. The aspartate residue at the N-terminal cleavage site is highlighted in red. (B,C) The RGF1 wild-type precursor (B) and its site-directed D104A mutant (C) were expressed and purified from E. coli. Precursor proteins were incubated with exudates from wild-type or *sbt3.8* mutant seedlings for the time indicated. Digests were separated on Tris-Tricine gels and stained by CBB. The cleavage product generated from the RGF1 precursor by wild-type exudates is indicated (arrow).

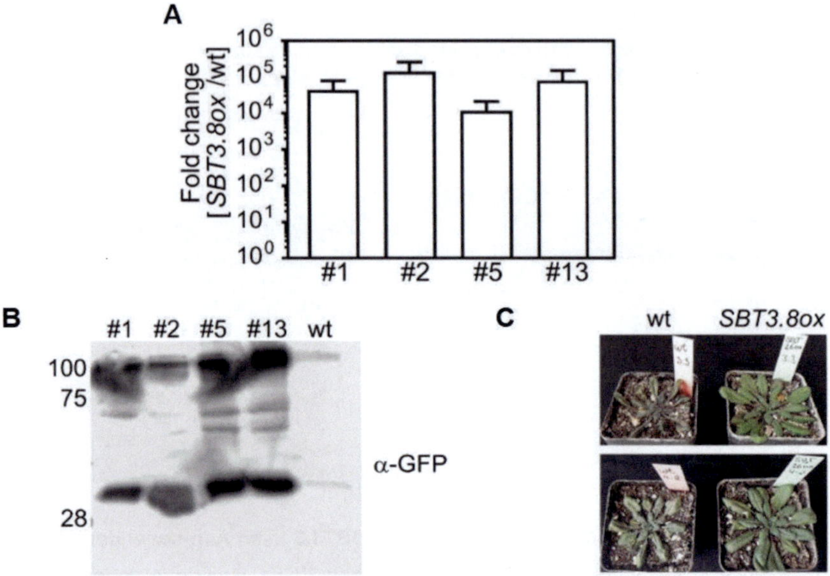

Supplementary Figure S6: Characterization of SBT3.8ox lines with respect to SBT3.8 transcript and protein levels, and drought stress tolerance. (A) Relative expression of SBT3.8 in four independent *P35S:SBT3.8-sfGFP* (SBT3.8ox) lines under normal growth conditions in comparison to the wild type (wt) is shown as fold change in SBT3.8ox lines vs wild-type plants. (B) SBT3.8-sfGFP protein levels in four independent SBT3.8ox lines in comparison to the wt. Leaf protein extracts (20 µg) were separated by SDS-PAGE and analyzed on anti-GFP immunoblots. (C) Representative wt and SBT3.8ox plants are shown that were grown for three weeks with sufficient water. Subsequently, plants were watered only every third day with 10 ml each. Pictures were taken after three weeks of water deficit.

2.4 A Quick Method to Analyze Peptide-Regulated Anthocyanin Biosynthesis

Eric Bühler[1], Andreas Schaller[1] and Nils Stührwohldt[1,#]

[1] Department of Plant Physiology and Biochemistry, Institute of Biology, University of Hohenheim, 70593 Stuttgart, Germany.

[#]correspondence: nils.stuehrwohldt@uni-hohenheim.de

Running head: Peptide-Regulated Anthocyanin Biosynthesis

Status: accepted manuscript, soon to be published in Methods in Molecular Biology as a book chapter of the "Plant Peptide Hormones and Growth Factors" volume, in 2024, by Springer Nature.

Own contribution: I contributed to the design of the protocol, to the writing and performed all experiments.

Summary/ Abstract

Post-translationally modified peptides are now recognized as important regulators of plant stress responses. We recently identified the sulfated CLE-LIKE6 (CLEL6) peptide as a negative regulator of anthocyanin biosynthesis in dark-grown and in light-stressed Arabidopsis seedlings. The function of CLEL6 depends on proteolytic processing by subtilisin-like serine proteinase SBT6.1, and on tyrosine sulfation by tyrosylprotein sulfotransferase (TPST), and CLEL6 signaling relies on the ROOT MERISTEM GROWTH FACTOR 1 INSENSITIVE (RGI) receptor family. In this chapter, we describe in detail how to quantify peptide-regulated and stress-induced anthocyanin biosynthesis. We include protocols for peptide treatment of Arabidopsis seedlings and growth under different stress conditions, for the extraction and quantification of anthocyanins, and for the expression analysis of anthocyanin biosynthetic genes.

Key words

anthocyanins, light-stress, stress response, CLEL peptides, gene expression, RNA extraction, cDNA synthesis, qPCR, photometric analysis

1 Introduction

Anthocyanins are secondary metabolites belonging to the class of flavonoids, and they are prominent pigments coloring flowers and fruits to attract pollinators as well as frugivore animals for seed dispersal. Over 600 different anthocyanins have been described that range in color from orange to red, blue and purple (Holton et al., 1995; Pervaiz et al., 2017). At the molecular level, anthocyanins are important stress metabolites produced in response to heat, cold, drought, light and nutrition stress. A common feature of these stresses is the accumulation of Reactive Oxygen Species (ROS). Due to their electron deficiency, anthocyanins are highly reactive against ROS, thereby protecting the plant from oxidative damage (Shao et al., 2007; Nakabayashi et al., 2014). Additional protection from light stress is provided by the fact that anthocyanins absorb photosynthetically active radiation, thereby protecting the photosynthetic apparatus from photoinhibition (Steyn et al., 2002; Xu et al., 2017).

Anthocyanin biosynthesis is controlled by many transcription factors including MYB11 for the regulation of early biosynthesis genes *CHALCONE SYNTHASE* (CHS), *CHALCONE ISOMERASE* (CHI), and *FLAVANONE 3-HYDROXYLASE*

(F3H), and PAP1 controlling *DIHYDROFLAVONOL 4-REDUCTASE* (DFR), *LEUCOANTHOCYANIDIN OXYGENASE* (LDOX) and *UDP-GLUCOSE:FLAVONOID 3-O-GLUCOSYLTRANSFERASE* (UF3GT) that code for enzymes late in the biosynthetic pathway (Yan et al., 2021). Light-dependent regulation is provided by the E3 (ubiquitin ligase) complex CONSTITUTIVE PHOTOMORPHOGENESIS1/SUPPRESSOR OF PHYA-105 (COP1/SPA), and the transcription factors PHYTOCHROME-INTERACTING FACTOR3 (PIF3) and ELONGATED HYPOCOTYL 5 (HY5) (Deng et al., 1992; Shin et al., 2007). Hormonal regulators of anthocyanin biosynthesis include ethylene, gibberellins, abscisic acid and jasmonates (Loreti et al., 2008). Furthermore, we recently identified the small post-translationally modified peptide CLE-LIKE6 (CLEL6) as a negative regulator of anthocyanin biosynthesis (Bühler et al., 2023). The activity of CLEL6 as an inhibitor of anthocyanin biosynthesis depends on post-translational tyrosine sulfation by tyrosylprotein sulfotransferase (TPST), proteolytic cleavage by the subtilisin-like serine proteinase SBT6.1, and perception by the ROOT MERISTEM GROWTH FACTOR 1 (RGI) receptor family (Bühler et al., 2023). CLEL6 down-regulates the expression of the MYB11 and PAP1 transcription factors and of anthocyanin biosynthesis genes to restrain anthocyanin formation (Bühler et al., 2023).

To assess the effect of signaling peptides on the plant stress response, we analyze anthocyanin biosynthesis and accumulation in peptide-treated vs. control seedlings. Here we include protocols for the treatment of Arabidopsis seedlings, the extraction of pigments and the photometric quantification of anthocyanins in seedling extracts. The effect of peptide treatment on anthocyanin biosynthesis is analyzed by reverse transcriptase (RT) qPCR. Protocols are included for the isolation of total RNA from peptide-treated and control seedlings, for cDNA synthesis and for gene expression analysis using oligonucleotide primers specific for anthocyanin biosynthesis genes and their transcriptional regulators. Finally, a comparison between the expression level of the genes involved and anthocyanin content yields important information on peptide-regulated anthocyanin biosynthesis.

2 Material

2.1 Growth of Plants

1. 94 mm petri dishes.
2. ½ MS Plates: Weigh in 1.075 g/L MS salts and 10 g/L sucrose. Dissolve in water. Add 3.8 g/L gelrite, make up to 1 L with water. Sterilize by autoclaving. Cool down to 60 °C. For peptide treatment, include 1 µM of the peptide from sterile stock solution. Poor into petri dishes. Store at 4 °C for max of one week. Optional for osmotic stress treatment: include 30 mM mannitol.
3. 70% (v/v) ethanol.
4. Sterile ddH_2O.
5. CLEL6 peptide (Sequence: DY(SO_3H)PQPHRKPPIHNE). Take up the lyophilized peptide in autoclaved ddH_2O to result in a 1 mM stock solution (*see* **Note 1**). Store at -20 °C. Avoid repeated freeze-thaw cycles.
6. Rotator.
7. Temperature-controlled plant growth cabinet.

2.2 Anthocyanin extraction and measurement

1. Bead beater, with 2 mL reaction tubes and steel beads. We recommend the bead beater for tissue homogenization. Alternatively, frozen tissue sample can be homogenized with mortar and pestle.
2. Liquid nitrogen.
3. Anthocyanin extraction solvent: 45 % (v/v) methanol, 5 % (v/v) acetic acid.
4. Photometer: We recommend to perform the assay in a 96-well plate format. This requires a microplate reader (e.g., TECAN SPARK). Alternatively, measure the samples in cuvettes using a conventional photometer.
5. 96-well microtiter plates (e.g., flat-bottom Greiner bio-one), or 1.6 mL disposable half micro cuvettes.
6. Microcentrifuge for 1.5 mL/2 mL microfuge tubes.

2.3 RNA extraction

1. RNAse free 1.5 mL reaction tubes. To avoid RNA degradation by RNAses we recommend using RNAse-free reaction tubes. Alternatively, you can use standard reaction tubes autoclaved for 45 min.
2. Micropestles (DNase/RNase-free) fitting the reaction tubes.

3. TRIzol reagent.
4. Chloroform.
5. Isopropanol.
6. 70 % (v/v) ethanol.
7. RNAse-free water.

2.4 cDNA synthesis

1. DNAseI (1 U/µL, e.g. Thermo Scientific)
2. 5x RT buffer: 250 mM Tris/HCl, pH 8.3, 250 mM KCl, 20 mM $MgCl_2$, 50 mM DTT
3. 25 mM EDTA: Weigh in 9.306 g/L Disodium EDTA dihydrate and dissolve in 80 % of the final volume of water (e.g., 800 mL for 1 L final volume). Adjust pH to 8.0 with NaOH while stirring until completely dissolved. Add water up to final volume. Store at room temperature.
4. 10 mM dNTPs.
5. 0.5 µg/µL Oligo(dT) primers.
6. Reverse Transcriptase (200 U/µL, e.g., RevertAid, Thermo Scientific)
7. Thermocycler.

2.5 qPCR analysis

1. qPCR Primers (*see* Table 1): obtain primers custom-synthesized a 0.01 µmol synthesis scale, purified and desalted. Take up in TE buffer (10 mM Tris/HCl, pH 8.0, 1 mM EDTA) to result in 100 µM stock solutions. Dilute in water to result in 1 µM working solutions. Store at -20 °C.

Table 1. qPCR primer list. ACTIN, TUB4 and CHS primers were previously published in (Bühler et al., 2023). F3'H, DFR, LDOX, UF3GT, UGT78D2, PAP1 and MYB11 primers were previously published in (Zhao et al., 2021).

Name	**Sequence** (5´-3´)
ACTIN2-F	GTGGATATCAGGAAGGATCTGTACGGT
ACTIN2-R	TTCTGTGAACGATTCCTGGACCTGC
Tub4-F	TGAGGAAGGAAGCTGAGAACAG
Tub4-F	ATACACTGGTCAGCGTTTTC
CHS-F	CCTGACTACTACTTCCGCATCACC
CHS-R	CGCACATGCGCTTGAACTTCTC
CHI-F	TCATGTAGACTCCGTCACGTTTG
CHI-R	TGACAGATAGAGAAGGAACGGCG
F3'H-F	CGGTGGACTGGGCTATAGCTGA
F3'H-R	CGAGAGTGGTGTTGGTGGATG
DFR-F	CTTTGTTCGTGCCACCGTTCG
DFR-R	AAAATCCATGGGTGTTGCCAC
LDOX-F	GTTTGCAGCTTTTCTACGAGGGC
LDOX -R	ATGTTGAGCAAAAGTCCGTGGAG
UF3GT-F	TTGTCAGATCGTTTTGGTTCCGC
UF3GT-R	TCTTCCTCACTTTCTCACCGATC
UGT78D2-F	CGGTGTTGGAGAGTGTATCGG
UGT78D2-R	CCAATCTCCCACACAACCTCC
PAP1-F	TGGTTCCTGAAGCGACGACAAC
PAP1-R	CGCAAACAAATGTTCGAAACAC
MYB11-F	GATGGCGATTGTAACCCAAGC
MYB11-R	ACATGAGGACACGTGGACAGC

2. 10000x SYBR Green solution in DMSO (e.g., SYBR Green I nucleic acid gel stain, Invitrogen). Dilute 1/7000 to obtain the SYBR Green working solution.
3. DNaseI (e.g. DNaseI, RNase-free, 1U/µL, ThermoFisher)
4. *Taq* DNA-polymerase: We use recombinant *Taq* DNA-polymerase expressed and purified in our lab. Alternatively, any of many commercial sources can be used.

5. 10 mM dNTPs.
6. 5x PCR puffer: 15 mM $MgCl_2$, 100 mM $(NH_4)_2SO_4$, 0.08 % (v/v) Triton X100, 10 % (v/v) DMSO, 250 mM KCl, 50 mM Tris/HCl, pH 8.3, 0.4 % (v/v) Tween 20.
7. Thermocycler.
8. Thermocycler for qPCR (e.g., CFX Connect Realtime System, Biorad).

3 Methods

3.1 Growth of Plants

1. Surface sterilization of seeds: Put approximately 60 seeds into a 1.5 mL reaction tube and add 1 mL of 70 % ethanol. Prepare one tube for each treatment you intend to do (e.g., control vs stress treatment, or control vs peptide treatment at different concentrations).
2. Incubate on rotator for 15 min at room temperature.
3. Perform steps 3 – 5 under a clean bench: Let the seeds settle to the bottom of the reaction tube and use a pipet to discard the supernatant ethanol.
4. Wash the seeds by adding 1 mL sterilized ddH_2O, let them settle again and remove the supernatant. Repeat this step two more times.
5. Pipet 45 seeds onto ½ MS plates, one tube per plate (see subheading 3.1, **step 1**) and distribute them evenly. We recommend pipetting the seeds in three lines, approximately 15 seeds each. Place them as far apart as possible.
6. Stratify seeds for two days at 4 °C.
7. Transfer the plates into a growth cabinet to grow seedlings for five days at 22 °C. For light stress experiments, keep them in darkness. Alternatively, grow seedlings at 100 µmol m^{-2} s^{-1} white light under (12h/12h) short-day conditions.
8. Induce stress to the Arabidopsis seedlings by specific treatments: for osmotic stress, transfer the seedlings on media supplemented with 300 mM mannitol. To induce light stress, transfer the plates with etiolated seedlings into continuous strong light (150 µmol m^{-2} s^{-1}). To induce heat stress, raise the temperature to 30 °C, or transfer the plates into a 30 °C growth cabinet at the same light regime. For peptide treatment see subheading 2.1, **step 2**.

Seedlings can either be grown on media including peptide, or be transferred on plates including peptide after five days.

3.2 Anthocyanin extraction and measurement

1. At the desired time point after stress treatment (*see* **Note 2**), collect plant material, about 30 entire seedlings as one sample, and determine the fresh weight (*see* **Note 3**).
2. Freeze samples in liquid nitrogen.
3. Homogenize the samples in a 2 mL reaction tube. Use either a bead beater or mortar and pestle pre-cooled with liquid nitrogen for tissue disruption.
4. Add 200 µL anthocyanin extraction solvent to each sample (*see* **Note 4**).
5. Mix thoroughly using a vortexer.
6. Centrifuge for 10 min at 13,000 x *g* and 4 °C.
7. Transfer supernatant into new reaction tube.
8. Repeat steps 6 and 7 to remove any insoluble remnants (*see* **Note 5**).
9. Transfer 100 µL of the supernatant into a 96 well plate for measurement in a microplate reader. Alternatively, add 100 µL of the supernatant to 900 µL of anthocyanin extraction solvent in a semi-micro cuvette for measurement in a conventional photometer (*see* **Note 6**).
10. Measure absorption at 534 nm and 657 nm for each sample. For calibration, use a blank including only the extraction solvent.
11. Calculate the anthocyanin content by using the following formula: $[Abs_{534}-(0.25 \times Abs_{657})] \times 2$ / g fresh weight.
12. Plot the results as shown in Fig. 1.

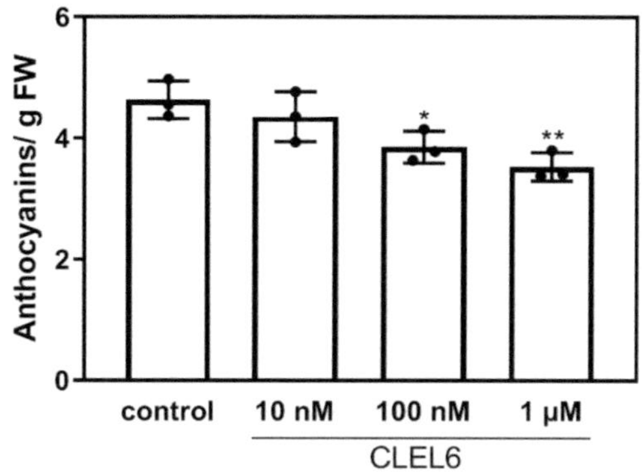

Fig. 1 CLEL6 treatment reduces the anthocyanin levels in wild-type seedlings in a dose-dependent manner. Quantification of the anthocyanin content in wild-type (wt) seedlings that were grown for five days in darkness followed by 16 h light stress (150 µmol m^{-2} s^{-1}). The growth medium was supplemented with 10 nM, 100 nM or 1 µM CLEL6 as indicated. Asterisks indicate significant differences (*P<0.05, **P<0.01; two-tailed t-test, n=3).

3.3 Gene expression analysis

Use RNAse free reaction tubes and pipet tips for all steps in this protocol (section 3.3).

3.3.1 RNA extraction

1. Collect all entire seedlings of one plate into one reaction tube and freeze them in liquid nitrogen.
2. Homogenize samples using either a bead beater or micropestles for tissue disruption (*see* **Note 7**).
3. Directly add 1 mL of TRIzol, mix thoroughly by vortexing and incubate at room temperature for 10 min.
4. Centrifugation for 10 min at 4 °C, 13000 x *g*.
5. Transfer supernatant into a new reaction tube.
6. Add 200 µL chloroform and incubate for 5 min at room temperature and shaking at 1500 rpm.
7. Centrifuge for 5 min at 4°C, 13000 x *g*.
8. Transfer upper aqueous phase into a new reaction tube.

9. Add 600 µL isopropanol and mix thoroughly.
10. Incubate at 4 °C for 30 min to precipitate RNA.
11. Centrifuge for 10 min at 4°C, 13000 x *g*.
12. Use a pipet to carefully discard the supernatant.
13. Add 1 mL of 70% ethanol and incubate for 5 min to wash the RNA pellet.
14. Centrifuge for 5 min at 4°C, 13000 x *g*.
15. Use a pipet to carefully remove as much of the supernatant as possible.
16. Dry the pellet by placing the open tubes under a clean bench for approximately 30 min.
17. Resuspend the pellet in 20 µl RNase-free water. Store at -20 °C.
18. Determine RNA concentration in a photometer. We recommend to use a photometer for very small volumes (e.g., Nanodrop, or TECAN microplate reader with NanoQuant Plate). Measure absorption at 260 nm and 280 nm. The Ratio 260/280 is a measure for the purity of the RNA. A value between 1.8 and 2 indicates purity, lower values result from contaminating proteins. Calculate RNA concentration by
C [µg/mL] = (OD260/pathlength) *dilution factor*40. The pathlength depends on the instrument used; for standard cuvettes it is 1 cm.

3.3.2 First strand cDNA synthesis

We recommend to perform the following steps in 0.5 mL PCR-reaction tubes in a thermocycler.

1. Use 1-10 µg of RNA (from subheading 3.3.1, **step 17**) and add RNase free water up to a final volume of 15 µL.
2. Add 4 µL of 5x RT-buffer and 1 µL (1U) DNaseI (*see* **Note 8**).
3. Incubate for 30 min at 37 °C.
4. Add 2 µl 25 mM EDTA.
5. Incubate for 15 min at 65 °C to inactivate DNaseI.
6. Add 2 µL of 0.5 µg/µL oligo(dT) primers.
7. Incubate for 5 min at 70 °C.
8. add 8 µL 5x RT-buffer, 4 µL dNTPs and 4 µL RNase-free water.
9. Incubate for 5 min at 37 °C.
10. Add 2 µL reverse transcriptase.
11. 11. Incubate for 1 h at 42 °C.

12. Incubate for 15 min at 70 °C to inactivate the reverse transcriptase. The resulting cDNA can be used directly for qPCR analysis, or stored at -20 °C. Depending on the amount of RNA used for cDNA synthesis, we recommend diluting the cDNA before qPCR analysis (e.g., dilute 1:10 for cDNA from 5 µg of RNA).

3.3.3 qPCR analysis

1. Prepare the following reaction mix for every primer combination (*see* **Notes 9 and 10**). We recommend to prepare a master mix containing all components but the cDNA template. It is highly recommended to perform triplicates for every sample of cDNA to avoid technical errors (*see* **Notes 11 and 12**).

   Table 2. Reaction mix for qPCR.

reagent	**volume [µL]**
ddH_2O	4.5
5x PCR buffer	4
dNTPs	0.5
Sybr Green	4
Primer F	2
Primer R	2
Taq Polymerase	1
cDNA	2
in total	**20**

2. Perform the qPCR with the following program:

   2 min at 95 °C
   10 sec at 95 °C |
   10 sec at 56 °C | x 44
   45 sec at 72 °C |

3. To check for the specificity of amplification include a melting curve at the end of the qPCR program (*see* **Note 13**).
4. When the qPCR is finished, check the Ct values for the reference genes (*ACTIN2* and *TUB4*) obtained for the different cDNA samples. They should

not exceed 35 (*see* **Note 14**) and they should not differ by more than 2 (*see* **Note 15**).

5. Calculate the treatment (peptide or stress)-induced change in relative expression of the target gene using the $2^{-\Delta\Delta Ct}$ method (Livak et al., 2001; Pfaffl 2001) (*see* **Note 16**). We recommend to use two reference ('house keeping') genes (*ACTIN2* and *TUB4*) for normalization. Calculate relative expression changes for both, and use the mean.
6. Plot the results as shown in Fig. 2.

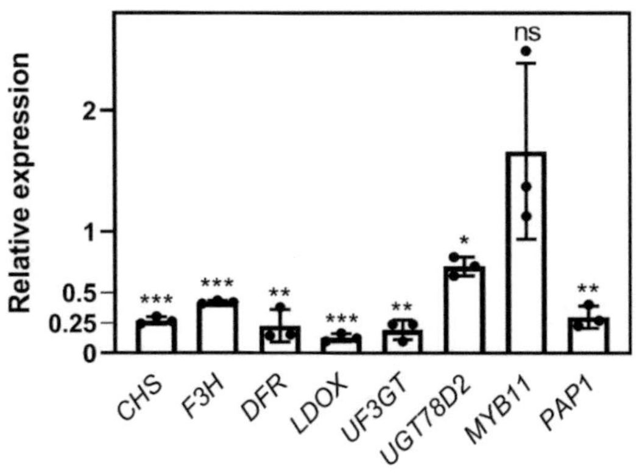

Fig. 2 CLEL6 treatment downregulates the expression of anthocyanin biosynthesis genes. Relative expression of *CHS*, *F3H*, *DFR*, *LDOX*, *UF3GT*, *UGT78D2*, *MYB11*, and *PAP1* in five-day-old etiolated seedlings treated with CLEL6 relative to the untreated control. 1 µM CLEL6 was added to the growth medium. Gene expression was normalized to *ACTIN2* and *TUB4* (n=3). ns indicates no significant difference; asterisks indicate significant differences at *P<0.05, **P<0.01, ***P<0.001 (one sample t-test).

4 Notes

1. Peptides should be custom-synthesized at > 90 % purity or higher.
2. We recommend 1 – 2 days of stress treatment before the sampling for anthocyanin measurement. For the analysis of stress-induced induction of anthocyanin biosynthesis genes, we recommend a shorter time of stress exposure of approximately 2 h before sampling. Alternatively, a time-course analysis can be performed.
3. For accurate fresh weight determination, use a tissue to cautiously clear plant samples of adhering water or medium.
4. The volume may have to be adjusted depending on the fresh weight of the samples. We recommend to use x µL per x mg of fresh weight. Be aware that you need to adapt the formula for the calculation of the anthocyanin content (subheading 3.2, **step 11**) if extraction volumes are changed.
5. It is very important that the anthocyanin extract is clear and does not contain any insoluble material that would affect photometric measurement. If it is not, repeat steps 6 and 7 as often as required.
6. We recommend to measure each sample in technical duplicates and calculate the mean.
7. Make sure that the plant material does not thaw during homogenization before you add the TRIzol reagent. Otherwise RNA will be degraded.
8. DNAseI is added to degrade any contaminating genomic DNA (Flohr et al., 2003). It is important that the product used is free of RNase activity.
9. In this protocol we included the primers recommended for the analysis of anthocyanin biosynthesis genes. If you want to analyze the expression of other genes, you will need different primers. For the design of primers, aim at an amplicon length of approximately 100-200 base pairs. The primers should be 20-25 nucleotides in length, and the melting points should be in the range of 55-63 °C. It is crucial that the primers bind specifically to the gene of interest and to no other targets in the genome. If possible, choose primers that span an exon-intron boundary to avoid the amplification of any residual genomic DNA. We recommend using a tool like QuantPrime (https://quantprime.mpimp-golm.mpg.de/) for primer design.
10. Primer efficiency needs to be determined experimentally before you use them for qRT-PCR analysis. To create a standard curve, prepare a serial

dilution of your cDNA (e.g. 0.5, 0.25, 0.125, 0.0625, 0,03125) and amplify these samples with the primer pair of interest in a standard qPCR. Plot the log of dilution against the measured Ct values and calculate the slope of regression. Calculate primer efficiency using the following formula:

$primer\ efficiency\ (\%) = (10^{-1/slope}-1)*100$

A pair of primers is suitable for qPCR analysis if efficiency is between 90 and 110 %. We recommend to use the Bio-Rad CFX Maestro tool to analyze and calculate primer efficiency easily.

11. Don't forget to include a non-template control (NTC) for every master mix. This can be used to evaluate contaminations.
12. To accurately compare different cDNA samples for a specific primer pair, they have to be amplified with the same master mix and run on the same qPCR plate.
13. The melting temperature depends on the length of the amplicon, which depends on the primer pair used. Therefore, the melting curve should be the same, with a specific peak for all samples amplified with the same pair of primers. If there is more than one peak, or no clear peak detectable, there are more than one PCR product, or primer dimers present. Additional amplicons may result from contamination with genomic DNA, or the amplification of off-targets. In this case the primers are not suitable and should be re-designed.
14. At Ct values above 35, results are not reliable. Repeat the analysis with an increased amount of cDNA template.
15. The Ct value of the house keeping genes (*ACTIN2* or *TUB4*) should not differ by more than 2 between the different samples. Otherwise they are not comparable. If the difference is too big, one of the cDNA samples should be diluted and the qPCR analysis repeated.
16. First calculate the relative expression of the reference gene (R; *ACTIN2* or *TUB4*) between treated and control sample. In this step the difference in the Ct of house keeping genes between the control and treated sample is calculated to later normalize the values of the target gene.

    $R=2^{\Delta Ct\ (Ct\ control\ sample-\ Ct\ treated\ sample)}$

    Second calculate the relative expression of the target gene (T). In this step the difference in the Ct of the target gene between the control and target sample is calculated. Thereby the fold change of the target gene between control and treated sample is calculated.

$T=2^{\Delta Ct\ (Ct\ control\ sample - Ct\ treated\ sample)}$

To normalize the fold change of the target gene, divide relative expression of target by that of the reference genes.

$N = T/R$

If primer efficiency ≠ 100%, modify the formulas for T and R by substituting '2' with 1+1*(primer efficiency(%))/100.

3. Discussion

In this study, the role of sulfated peptides in regulating stress and development responses in *Arabidopsis thaliana* was investigated. It was observed that PSK plays a crucial role in enhancing drought resistance, while CLEL6 is involved in regulating the accumulation of anthocyanins under light stress conditions. Furthermore, sulfated peptides were found to be associated with the transition from skoto- to photomorphogenic development.

Specific points regarding these findings were addressed in the individual discussion sections in the results (chapter 2). The forthcoming discussion will present a broader perspective on the role of sulfated peptides in stress responses and development.

3.1 Sulfated peptides as abiotic stress regulators

In this work, we identified the sulfated peptides PSK and CLEL6 as abiotic stress regulators, which they now share with the sulfopeptide family of PSYs (Ogawa-Ohnishi et al., 2022). These peptides belong to the class of small posttranslational modified peptides, which are known regulators of plant stress responses. Many have been identified in the family of CLAVATA3(CLV)/EMBRYO-SURROUNDING REGION RELATED (CLE). For example, during drought, CLE25 induces the expression of the *NCED3* gene, which encodes an enzyme involved in the rate-limiting step of ABA synthesis, while CLE9 induces stomatal closure (Takahashi et al., 2018; Zhang et al., 2018). CLE14 regulates stress-induced leaf senescence by modulating ROS homeostasis (Zhang et al., 2021), and CLE45 promotes pollen tube growth under heat stress (Endo et al., 2013). Furthermore, IDA regulates leaf abscission upon drought (Patharkar et al., 2016), and the recently identified IDA-like (IDL) peptides, IDL6 and IDL7, negatively regulate the ROS-mediated stress responses (Vie et al., 2017). From the family of C-terminally encoded peptides, CEP5 regulates osmotic stress tolerance (Smith et al., 2020), while CEP1 acts as a shoot-to-root signal during N starvation to adjust nitrogen acquisition (Ohkubo et al., 2017). RALF22 and RALF23 regulate salt stress-induced cell wall changes, thereby influencing plant growth and salt tolerance (Zhao et al., 2021). Additionally, the PLANT ELICITOR PEPTIDE3 (PEP3) has been shown to regulate salinity stress (Nakaminami et al., 2018). Overall, the findings of this study contribute to the understanding of sulfated peptides as part of a larger group of small posttranslationally modified peptides involved in the regulation of abiotic stress responses.

3.2 Sulfated peptides as biotic stress regulators

Peptide signals play a crucial role in regulating plant responses to biotic stress. In tomato (*Solanum lycopersicum*), the release of the systemin peptide upon wounding triggers the systemic immune response (Pearce et al., 1991; Wang et al 2018). Similarly, PATHOGENESIS-RELATED PROTEIN1b (PR-1b) the precursor for CAP-DERIVED PEPTIDE1 (CAPE1) activates immune responses (Chen et al., 2014). In Arabidopsis, the genes encoding PEPs are upregulated during biotic stresses, and the PEPs themselves act as DAMAGE ASSOCIATED MOLECULAR PATTERNS (DAMPs), triggering plant immunity through the PEPR1 and PEPR2 LRR-RLKs (Bartels et al., 2015). The secreted PATHOGEN-ASSOCIATED MOLECULAR PATTERN (PAMP)-INDUCED PEPTIDES (PIPs) are perceived by RECEPTOR-LIKE KINASE7 (RLK7) and amplify the immune response (Hou et al., 2014). SERINE-RICH ENDOGENOUS PEPTIDES (SCOOPs) are expressed in response to herbivory and contribute to resistance through the MIK2 receptor (Stahl et al., 2022). The precursor of RALF23 peptide is cleaved by SBT6.1 and inhibits the plant immune response by binding to the FERONIA receptor (Stegmann et al., 2017).

Within the sub-group of sulfated peptides, PSK has been implicated in pathogen interactions. Mutant plants lacking PSK receptors (*pskr1-3*) exhibit reduced crown gall formation upon *Agrobacterium tumefaciens* infection (Loivamäki et al., 2010). Furthermore, PSK antagonistically regulates defense responses to biotrophic and necrotrophic pathogens through its receptor PSKR1, as the *pskr1* mutant showed reduced resistance to fungi, while bacterial resistance was increased (Mosher et al., 2012). Additionally, susceptibility to oomycetes and nematodes is impaired in the *pskr1* mutant, indicating an important role for PSK in pattern-triggered immunity (PTI) against a broad range of pathogens (Rodiuc et al., 2016). The nine PSY peptides redundantly mediate the trade-off between plant growth and stress response in abiotic and biotic stress (Ogawa-Ohnishi et al., 2022). *Xanthomonas oryzae* produces a sulfated peptide that mimics the PSYs, thereby exploiting their function and increasing its virulence to plants (Pruitt et al., 2017). Within the family of RGFs, the after bacterial infection highly upregulated RGF7 acts as a DAMP to regulate plant immunity through the RGI4/RGI5 receptors using SERK4/BAK1 as co-receptors (Wang et al., 2020). RGF9 binds to the RGI3 receptor, which subsequently interacts with the pattern-triggered receptor FLS2, thus raising the cell surface immune receptor abundance and regulating the immune response in Arabidopsis (Stegmann et al., 2022). The RGF1-RGI1 module regulates growth upon PTI (Liu et al., 2022). However, the PTI-triggered alkalinization of the acidic extracellular pH in the apical meristem inhibits the sulfotyrosine-dependent interaction of RGF1 with RGI1, thereby classifying sulfotyrosine as a pH sensor (Liu et al., 2022). Additionally, alkalinization of the apoplast was reported in response to biotic stress as well as abiotic stresses like salinity and drought (Geilfus 2017). In fact, pH dependent peptide perception by receptors occurs also for CLE-peptides, since Arabidopsis root can only sense CLE45 under acidic apoplast pH, although CLE45 lacks tyrosine sulfation (Diaz-Ardila et al., 2023; Han et al., 2023). Thus, the question arises if the perception of other RGFs (e.g. CLELs) use a similar pH dependent mechanism. Overall, all RGI family members except RGI2 are involved in the biotic stress response, raising the possibility that additional RGFs may also regulate biotic stress responses. RGF6 (also known as CLEL6) is a promising candidate, as overexpression of CLEL6 results in enhanced flg22-triggered PTI against bacterial pathogens, similar to the effect observed with CLEL9 overexpression (Stegmann et al., 2022).

3.3 Unidentified role of CLEL9 in light signaling

In this study, we observed a remarkable upregulation of *CLEL9* (*RGF9*) expression under continuous light (Result section 2.1, Supplemental figure S3). Given the fact that CLEL9 did not affect anthocyanin production, this indicates an unidentified role for CLEL9 in light signaling. RNA sequencing data from two independent studies underline this hypothesis, since *CLEL9* expression was significantly higher in the *pifq* mutant (lacking PIF1, PIF3, PIF4 and PIF5) compared to the wt (Leivar et al., 2009; Martin et al., 2016). Furthermore, comparing *pifq* to the triple mutant *pif1pif3pif4*, which lacks the PIF5 mutation, the *CLEL9* expression is still upregulated, similar to the comparison of *pifq*/wt, revealing a potential inhibitory role for PIF5 in regulation of *CLEL9*. Subsequently, promoter binding prediction analysis identified *CLEL9* as a direct target of PIF5 (Leivar et al., 2009). qPCR analysis of an overexpressing line of PIF5 (*35S:PIF5*) provided further support, as *CLEL9* transcripts were significantly less abundant compared to the wild type (Figure 10).

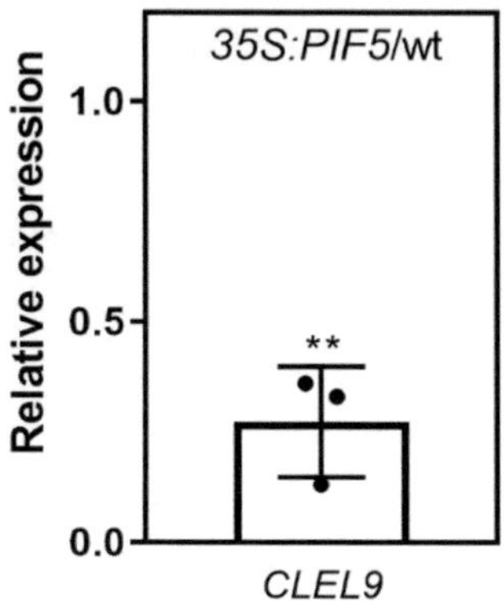

Figure 10. *CLEL9* expression lowered in *35S:PIF5* seedlings. Relative Expression of *CLEL9* in five-day-old etiolated *35S:PIF5* seedlings compared to the wt control. Seedlings were exposed to light for 2 h before sampling. Gene expression was normalized to two reference genes. Data are shown as the mean ±SD of n=3 independent experiments. Asterisks indicate significant differences (***P*<0.005; one sample *t*-test). Experimental setup including plant growth and qPCR was performed as described in the method section of results 2.1.

Apart from the PIF typical induction of hypocotyl growth, PIF5 is known to regulate various light responses. It is involved in the response to shade (Lorrain et al., 2008), induces leaf senescence (Sakuraba et al., 2014), regulates the phototropic response (Sun et al., 2014), and is associated with dark-induced senescence and chlorophyll degradation (Zhang et al., 2015). Given that CLEL9 has already been implicated in the regulation of seedling gravitropism, it is conceivable that it may also play a role in the phototropic response. Considering our data indicating the regulation of CLEL9 by PIF5, further investigation is needed to uncover the specific role of CLEL9 in light signaling, particularly in the context of PIF5 signaling.

3.4 An unknown sulfated peptide contributes to photomorphogenic development

This section will discuss the observed early photomorphogenesis phenotype of *tpst* mutants, which was described in this work and in two recent studies (Giorgi et al., 2021; Kang et al., 2014). Kang et al. (2014) initially described the early photomorphogenesis phenotype of *tpst-2* under phosphate deficiency and assumed the GLK transcription factors to be involved. This assumption was confirmed in this work. On the other hand, Giorgi et al. (2021) reported that the early photomorphogenesis is due to a defect in the *tpst* seedling cuticle, which is caused by non-sulfated and thus inactive PSY1 and CIF2.

Although the study proves that the early photomorphogenesis phenotype can be triggered by a defect in the seedling cuticle, there is no assumption about how this is mediated. Additionally, the early photomorphogenesis phenotype of the *psy1/cif1/cif2* mutant is weaker than that of *tpst*, indicating the involvement of other sulfopeptides in this mechanism. Due to the highly pleiotropic phenotypes of *tpst* mutants, speculation about the missing sulfopeptide is complex. However, there are two possibilities for how the unknown sulfopeptide functions: either it is involved in endosperm development like PSY1 and CIF2 and triggers the early photomorphogenesis through a developmental defect, or it directly regulates photomorphogenesis at a later time point of seedling development. In the first case, RGF7 would be a promising candidate as it exhibits the strongest expression in the endosperm among all sulfopeptides (Giorgi et al., 2021). Investigating this hypothesis could involve creating a combined *rgf7/psy1/cif2* mutant. In the second case, where the unknown sulfopeptide directly regulates photomorphogenesis, the peptide should be light-regulated. CLEL6 and CLEL9 have already been shown to be regulated by light and are therefore valid candidates. Speculating along these lines, further analysis of RNA-seq data from *pifq* mutants (Oh et al., 2012) revealed downregulation of *PSK3* and *PSK4* compared to the wild type. Considering that PSK, CLEL6, and CLEL9 can regulate hypocotyl length (Stührwohldt et al., 2011; Ghorbani et al., 2016), all three should be investigated for potential roles in the regulation of transition from skoto- to photomorphogenic development.

3.5 Sulfated peptides contribute to thermomorphogenesis

During the research of this work, *TPST* and *CLEL6* were identified as potential targets of the transcription factor PIF4 (Oh et al., 2012). PIF4 has been recognized as a key regulator of thermomorphogenesis, which includes all the morphological changes in plants triggered by elevated temperatures (Quint et al., 2016). These changes include early flowering, reduced stomatal density, elongated hypocotyls and petioles, and leaf hyponasty, which helps increase the distance between the plant and the heat-radiating ground to facilitate optimal ventilation and cooling (Quint et al., 2016; Casal et al., 2019; Lee et al., 2021). The crucial role of PIF4 in thermomorphogenesis has been validated by the observation that *pif4* mutants lack hypocotyl elongation, hyponasty, and petiole elongation in response to elevated temperatures (Figure 10). The regulation of PIF4 occurs at both the gene expression and protein abundance level. The transcription factor ELF3, as part of the evening complex, restricts the expression of PIF4. However, this inhibition is weakened under elevated temperatures (Nurisov et al., 2011). The abundance of PIF4 protein is controlled by the thermosensor phytochrome B (phyB). PhyB has been identified as a thermosensor due to its thermoresponsive dark reversion from the active Pfr form to the inactive Pr form (Legris et al., 2016). Dark reversion accelerates at elevated temperature (Jung et al., 2016). Active phyB promotes the degradation of PIF4, thus a faster dark reversion leads to increased PIF4 abundance.

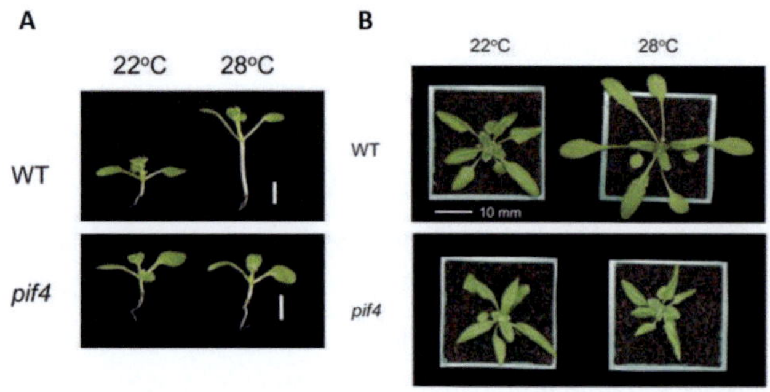

Figure 11. PIF4 is the key regulator of thermomorphogenesis. A Arabidopsis seedlings of WT (Col0) and *pif4* were grown four days at 22 °C and three additional days at 28 °C. Controls were kept seven days at 22 °C. **B** Arabidopsis plants of WT (Col0) and *pif4* were grown two weeks at 22 °C and three additional weeks at 28 °C. Controls were kept five weeks at 22 °C. Adapted from Koini et al., 2009 and reprinted with permission from Elsevier.

Various phytohormones contribute to adaptation to elevated temperatures including auxin which acts downstream of PIF4 (Lu et al., 2021). PIF4 directly activates the expression of *YUCCA8*, which enhances auxin synthesis and facilitates growth adaptation (Sun et al., 2012). *YUCCA8* encodes the indole-3-pyruvate monooxygenase involved in indole-3-acetic acid (IAA) synthesis. Brassinosteroids, stimulate growth via the transcription factor BRASSINAZOLE RESISTANT 1 (BZR1), which accumulates in the nucleus in response to high temperatures. Additionally, BZR1 promotes a feedforward loop by activating the *PIF4* expression (Ibañez et al., 2018). The PIF4-induced transcription factor TCP15 increases the expression of genes involved in gibberellin synthesis, thereby modulating hypocotyl growth (Ferrero et al., 2019). Elevated temperature-induced reduction in bioactive jasmonate levels leads to the stabilization of JASMONATE ZIM-DOMAIN (JAZ) proteins, which promote hypocotyl elongation (Zhu et al., 2021; Li et al., 2021). Although the involvement of peptide hormones in thermomorphogenesis has not been established so far, considering other abiotic or biotic stresses, it is feasible that sulfopeptides may play a role in thermomorphogenesis. The predicted regulation of the *TPST* gene by PIF4 is in line with this proposition. Our unpublished data support this hypothesis, as the increased hypocotyl growth observed in *tpst* mutants under elevated temperature treatment is significantly lower compared to the wild type (Figure 12).

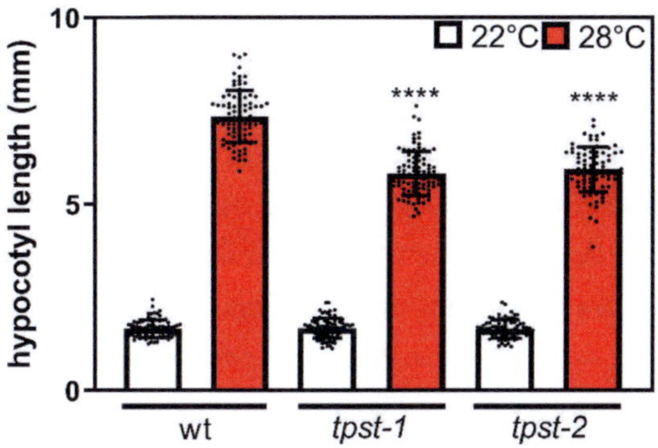

Figure 12. *tpst*-mutants display a reduced thermomorphogenesis phenotype. Wild-type (wt), *tpst-1* and *tpst-2* were grown for five days in 22°C. Aftewards they were grown four additional days in 22°C or 28°C respectively. n=3 with a total of at least 79 seedlings measured per line. Asterisks indicate significant differences against wt 28°C (****P<0.0001; two-tailed t-test). Experimental setup including plant growth and hypocotyl measurement was performed as described in the method section of results 2.1.

By exploring the dataset of predicted PIF4 targets, additional genes encoding sulfopeptides, such as CLEL6, RGF5, PSK2, PSK3, PSK4, and PSY1, were identified (Oh et al., 2012). CLEL6 (Ghorbani et al., 2016), PSK (Stührwohldt et al., 2011), and PSY1 (Fuglsang et al., 2014) have all been shown to regulate hypocotyl growth, suggesting their potential involvement in temperature-induced growth responses. Recently, it was demonstrated that the Arabidopsis root can sense and respond to elevated temperatures, making it feasible for root-expressed sulfopeptides such as RGF5, PSK, or PSY1 to contribute to this mechanism (Delker et al., 2022). Future studies are needed to provide further insights into the potential role of sulfopeptides in thermomorphogenesis.

3.6 Sulfated peptides are present in various species

Vast of the previous research, including the present study, addressed the role of sulfopeptides regulation of growth, development, and stress responses in Arabidopsis. The extent to which the funtions of these peptides is conserved across other species is largely unknown. However, sulfopeptides appear to be widespread in land plants. Firstly, the *TPST* gene, which is essential for sulfopeptide formation, is conserved across various land plants, including angiosperms, gymnosperms, mosses, liverworts, hornworts, ferns, and lycophytes (Furumizu et al., 2021a). CIFs are present in angiosperms, gymnosperms, liverworts, and hornworts, with the exception of TWS1 (Figure 13). TWS1 is only found in the seed producing angio- and gymnosperms, which is consistent with its regulatory role in seed development. The function of CIFs in liverworts and hornworts is yet to be discovered, as in Arabidopsis they regulate Casparian strip formation and pollen wall development. The PSY peptide family appears in angiosperms, gymnosperms, ferns, and mosses (Figure 13). The RGFs are conserved within vascular plants (angiosperms, gymnosperms, ferns, lycophytes) and liverworts (Figure 13), suggesting that RGF sequences were lost in mosses and hornworts. Their presence in rootless liverworts indicates potential functions of RGFs beyond root development, which would be interesting to investigate.

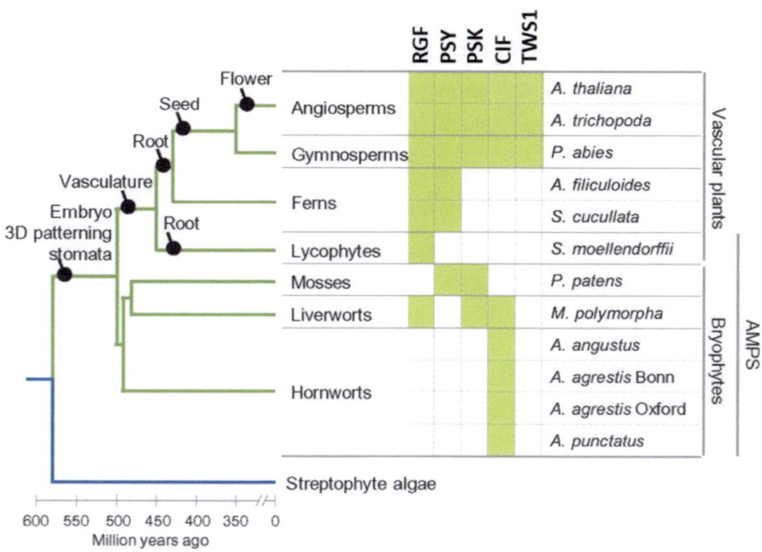

Figure 13. Conservation of sulfopeptides across species. The simplified phylogenetic tree of prominent extant land plant lineages (indicated by green lines) is derived from (One Thousand Plant Transcriptomes Initiative). Morphological treats are marked as black dots. Green boxes indicate the occurrence of a sulfopeptide in the species. Graph was adapted from Furumizu et al., 2021a.

Two RGF sequences have been reported in *Marchantia polymorpha,* 11 in *Solanum lycopersicum*, and 18 in *Zea mays* (Furumizu et al., 2021a). Selected members of the RGF family, such as Marchantia RGF1-peptide (DYAEPDTHPPESN), tomato RGF1 (DYKSPRHHPPRHN), and maize RGF1 (DYHSVHRHPPTHN), where expressed or added to Arabidopsis seedlings and caused similar wavy root phenotypes as Arabidopsis RGF1 (Fang et al., 2021; Meng et al., 2012). These non-Arabidopsis RGFs exhibited conservation of aspartate, tyrosine at the N-term, proline, and asparagine at the C-terminus, but have significant variation in other amino acid residues. Nonetheless they are perceived by the RGI receptor family, as the quadruple receptor mutant *rgi1234* was insensitive and did not show wavy roots in response to peptide treatment (Fang et al., 2021). Additionally, overexpression of MpRGF1 in Marchantia caused dwarfism, suggesting a growth-regulating role of MpRGF1 (Furumizu et al., 2021b). As the functions of the Arabidopsis RGF family extends beyond root-related processes, further studies should aim to elucidate non-root-related functions in other species.

Discussion

PSK is conserved in angiosperms, gymnosperms, mosses, and liverworts (Figure 13). The biological functions of PSK peptides in species other than Arabidopsis has been investigated in various studies. In woody plants like *Cunninghamia lanceolata*, PSK promotes somatic embryogenesis (Hao et al., 2023). A similar promoting effect on somatic embryogenesis was observed in carrot (*Daucus carota*) (Hanai et al., 2000). While studying the effect of PSK, other homologs were revealed. Recently, a new PSK-γ peptide with the sequence YVYTQ was found in soybean. Overexpression of this peptide in Arabidopsis resulted in increased growth of vegetative organs and seeds but did not influence PTI against bacterial pathogens (Liu et al., 2019; Yu et al., 2019). The legume-specific PSK-δ peptide with the sequence YIYTN regulates nodule organogenesis (Yu et al., 2022). PSK-ε (YVYTN) regulates root elongation and the formation of lateral roots in *Medicago truncatula* (Di et al., 2022).

In conclusion, sulfopeptides are important regulators of plant stress responses and due to their conservation across land plants, they might fulfil their functions in many species. This project could successfully provide further knowledge about how specific sulfopeptides contribute to the stress regulation in Arabidopsis. CLEL6 inhibits the light stress induction of anthocyanin biosynthesis, PSK mediates drought stress tolerance and sulfopeptides contribute to regulation of photomorphogenesis, while the specific peptides in this process remain unknown. Although this study contributes to the understanding of how sulfopeptides operate, more functions especially in other species than Arabidopsis, are yet to be discovered.

4. References

One thousand plant transcriptomes and the phylogenomics of green plants (2019). In: *Nature* 574 (7780), S. 679–685. DOI: 10.1038/s41586-019-1693-2.

Aghdam, Morteza Soleimani; Alikhani-Koupaei, Majid (2021a): Exogenous phytosulfokine α (PSKα) applying delays senescence and relief decay in strawberry fruits during cold storage by sufficient intracellular ATP and NADPH availability. In: *Food Chemistry* 336, S. 127685. DOI: 10.1016/j.foodchem.2020.127685.

Aghdam, Morteza Soleimani; Ebrahimi, Amin; Sheikh-Assadi, Morteza (2021b): Phytosulfokine α (PSKα) delays senescence and reinforces SUMO1/SUMO E3 ligase SIZ1 signaling pathway in cut rose flowers (Rosa hybrida cv. Angelina). In: *Scientific Reports* 11 (1), S. 23227. DOI: 10.1038/s41598-021-02712-2.

Amano, Yukari; Tsubouchi, Hiroko; Shinohara, Hidefumi; Ogawa, Mari; Matsubayashi, Yoshikatsu (2007): Tyrosine-sulfated glycopeptide involved in cellular proliferation and expansion in Arabidopsis. In: *Proceedings of the National Academy of Sciences of the United States of America* 104 (46), S. 18333–18338. DOI: 10.1073/pnas.0706403104.

An, Zhichao; Liu, Yuliang; Ou, Yang; Li, Jia; Zhang, Baowen; Sun, Daye et al. (2018): Regulation of the stability of RGF1 receptor by the ubiquitin-specific proteases UBP12/UBP13 is critical for root meristem maintenance. In: *Proceedings of the National Academy of Sciences of the United States of America* 115 (5), S. 1123–1128. DOI: 10.1073/pnas.1714177115.

Annunen, P.; Autio-Harmainen, H.; Kivirikko, K. I. (1998): The novel type II prolyl 4-hydroxylase is the main enzyme form in chondrocytes and capillary endothelial cells, whereas the type I enzyme predominates in most cells. In: *The Journal of Biological Chemistry* 273 (11), S. 5989–5992. DOI: 10.1074/jbc.273.11.5989.

Armstrong, G. A.; Runge, S.; Frick, G.; Sperling, U.; Apel, K. (1995): Identification of NADPH:protochlorophyllide oxidoreductases A and B: a branched pathway for light-dependent chlorophyll biosynthesis in Arabidopsis thaliana. In: *Plant Physiology* 108 (4), S. 1505–1517. DOI: 10.1104/pp.108.4.1505.

References

Bari, Rajendra; Jones, Jonathan D. G. (2009): Role of plant hormones in plant defence responses. In: *Plant Molecular Biology* 69 (4), S. 473–488. DOI: 10.1007/s11103-008-9435-0.

Bartels, Sebastian; Boller, Thomas (2015): Quo vadis, Pep? Plant elicitor peptides at the crossroads of immunity, stress, and development. In: *Journal of Experimental Botany* 66 (17), S. 5183–5193. DOI: 10.1093/jxb/erv180.

Bauer, Diana; Viczián, András; Kircher, Stefan; Nobis, Tabea; Nitschke, Roland; Kunkel, Tim et al. (2004): Constitutive photomorphogenesis 1 and multiple photoreceptors control degradation of phytochrome interacting factor 3, a transcription factor required for light signaling in Arabidopsis. In: *The Plant Cell* 16 (6), S. 1433–1445. DOI: 10.1105/tpc.021568.

Bellaire, Anke; Ischebeck, Till; Staedler, Yannick; Weinhaeuser, Isabell; Mair, Andrea; Parameswaran, Sriram et al. (2014): Metabolism and development - integration of micro computed tomography data and metabolite profiling reveals metabolic reprogramming from floral initiation to silique development. In: *New Phytologist* 202 (1), S. 322–335. DOI: 10.1111/nph.12631.

Beloshistov, Roman E.; Dreizler, Konrad; Galiullina, Raisa A.; Tuzhikov, Alexander I.; Serebryakova, Marina V.; Reichardt, Sven et al. (2018): Phytaspase-mediated precursor processing and maturation of the wound hormone systemin. In: *New Phytologist* 218 (3), S. 1167–1178. DOI: 10.1111/nph.14568.

Bharath, Pulimamidi; Gahir, Shashibhushan; Raghavendra, Agepati S. (2021): Abscisic acid-induced stomatal closure: An important component of plant defense against abiotic and biotic stress. In: *Frontiers in Plant Science* 12, S. 615114. DOI: 10.3389/fpls.2021.615114.

Brandt, Benjamin; Hothorn, Michael (2016): SERK co-receptor kinases. In: *Current Biology* 26 (6), R225-6. DOI: 10.1016/j.cub.2015.12.014.

Bühler, Eric; Fahrbach, Elisa; Schaller, Andreas; Stührwohldt, Nils (2023): Sulfopeptide CLEL6 inhibits anthocyanin biosynthesis in Arabidopsis thaliana. In: *Plant Physiology*. DOI: 10.1093/plphys/kiad316.

Carrington, J. C.; Freed, D. D.; Oh, C. S. (1990): Expression of potyviral polyproteins in transgenic plants reveals three proteolytic activities required for complete processing. In: *The EMBO Journal* 9 (5), S. 1347–1353. DOI: 10.1002/j.1460-2075.1990.tb08249.x.

Casal, Jorge J.; Balasubramanian, Sureshkumar (2019): Thermomorphogenesis. In: *Annual Review of Plant Biology* 70, S. 321–346. DOI: 10.1146/annurev-arplant-050718-095919.

Caspar, T.; Lin, T. P.; Kakefuda, G.; Benbow, L.; Preiss, J.; Somerville, C. (1991): Mutants of Arabidopsis with altered regulation of starch degradation. In: *Plant Physiology* 95 (4), S. 1181–1188. DOI: 10.1104/pp.95.4.1181.

Chang, Jinke; Li, Xiaopeng; Fu, Weihao; Wang, Jiawen; Yong, Yueyuan; Shi, Hongyong et al. (2019): Asymmetric distribution of cytokinins determines root hydrotropism in Arabidopsis thaliana. In: *Cell Research* 29 (12), S. 984–993. DOI: 10.1038/s41422-019-0239-3.

Chen, Y. F.; Matsubayashi, Y.; Sakagami, Y. (2000): Peptide growth factor phytosulfokine-alpha contributes to the pollen population effect. In: *Planta* 211 (5), S. 752–755. DOI: 10.1007/s004250000370.

Chen, Ying-Lan; Lee, Chi-Ying; Cheng, Kai-Tan; Chang, Wei-Hung; Huang, Rong-Nan; Nam, Hong Gil; Chen, Yet-Ran (2014): Quantitative peptidomics study reveals that a wound-induced peptide from PR-1 regulates immune signaling in tomato. In: *The Plant Cell* 26 (10), S. 4135–4148. DOI: 10.1105/tpc.114.131185.

Chichkova, Nina V.; Galiullina, Raisa A.; Mochalova, Larisa V.; Trusova, Svetlana V.; Sobri, Zulfazli M.; Gallois, Patrick; Vartapetian, Andrey B. (2018): Arabidopsis thaliana phytaspase: identification and peculiar properties. In: *Functional Plant Biology* 45 (2), S. 171–179. DOI: 10.1071/FP16321.

Chien, Pei-Shan; Nam, Hong Gil; Chen, Yet-Ran (2015): A salt-regulated peptide derived from the CAP superfamily protein negatively regulates salt-stress tolerance in Arabidopsis. In: *Journal of Experimental Botany* 66 (17), S. 5301–5313. DOI: 10.1093/jxb/erv263.

Chinchilla, Delphine; Zipfel, Cyril; Robatzek, Silke; Kemmerling, Birgit; Nürnberger, Thorsten; Jones, Jonathan D. G. et al. (2007): A flagellin-induced complex of the receptor FLS2 and BAK1 initiates plant defence. In: *Nature* 448 (7152), S. 497–500. DOI: 10.1038/nature05999.

References

Cho, Sung Ki; Larue, Clayton T.; Chevalier, David; Wang, Huachun; Jinn, Tsung-Luo; Zhang, Shuqun; Walker, John C. (2008): Regulation of floral organ abscission in Arabidopsis thaliana. In: *Proceedings of the National Academy of Sciences of the United States of America* 105 (40), S. 15629–15634. DOI: 10.1073/pnas.0805539105.

Clough, S. J.; Bent, A. F. (1998): Floral dip: a simplified method for Agrobacterium-mediated transformation of Arabidopsis thaliana. In: *The Plant Journal* 16 (6), S. 735–743. DOI: 10.1046/j.1365-313x.1998.00343.x.

Crevillén, Pedro; Ballicora, Miguel A.; Mérida, Angel; Preiss, Jack; Romero, José M. (2003): The different large subunit isoforms of Arabidopsis thaliana ADP-glucose pyrophosphorylase confer distinct kinetic and regulatory properties to the heterotetrameric enzyme. In: *The Journal of Biological Chemistry* 278 (31), S. 28508–28515. DOI: 10.1074/jbc.M304280200.

Delker, Carolin; Quint, Marcel; Wigge, Philip A. (2022): Recent advances in understanding thermomorphogenesis signaling. In: *Current Opinion in Plant Biology* 68, S. 102231. DOI: 10.1016/j.pbi.2022.102231.

Deng, X. W.; Caspar, T.; Quail, P. H. (1991): *cop1*: a regulatory locus involved in light-controlled development and gene expression in Arabidopsis. In: *Genes & Development* 5 (7), S. 1172–1182. DOI: 10.1101/gad.5.7.1172.

Deng, X. W.; Matsui, M.; Wei, N.; Wagner, D.; Chu, A. M.; Feldmann, K. A.; Quail, P. H. (1992): COP1, an Arabidopsis regulatory gene, encodes a protein with both a zinc-binding motif and a G beta homologous domain. In: *Cell* 71 (5), S. 791–801. DOI: 10.1016/0092-8674(92)90555-q.

Di, Qi; Li, Yi; Zhang, Danping; Wu, Wei; Zhang, Lin; Zhao, Xing et al. (2022): A novel type of phytosulfokine, PSK-ε, positively regulates root elongation and formation of lateral roots and root nodules in Medicago truncatula. In: *Plant Signaling & Behavior* 17 (1), S. 2134672. DOI: 10.1080/15592324.2022.2134672.

Diaz-Ardila, H. Nicholay; Gujas, Bojan; Wang, Qian; Moret, Bernard; Hardtke, Christian S. (2023): pH-dependent CLE peptide perception permits phloem differentiation in Arabidopsis roots. In: *Current Biology* 33 (3), 597-605.e3. DOI: 10.1016/j.cub.2022.12.056.

Ding, Shuting; Lv, Jianrong; Hu, Zhangjian; Wang, Jiao; Wang, Ping; Yu, Jingquan et al. (2023): Phytosulfokine peptide optimizes plant growth and defense via glutamine synthetase GS2 phosphorylation in tomato. In: *The EMBO Journal* 42 (6), e111858. DOI: 10.15252/embj.2022111858.

Doblas, Verónica G.; Smakowska-Luzan, Elwira; Fujita, Satoshi; Alassimone, Julien; Barberon, Marie; Madalinski, Mathias et al. (2017): Root diffusion barrier control by a vasculature-derived peptide binding to the SGN3 receptor. In: *Science* 355 (6322), S. 280–284. DOI: 10.1126/science.aaj1562.

Doll, N. M.; Royek, S.; Fujita, S.; Okuda, S.; Chamot, S.; Stintzi, A. et al. (2020): A two-way molecular dialogue between embryo and endosperm is required for seed development. In: *Science* 367 (6476), S. 431–435. DOI: 10.1126/science.aaz4131.

Duek, Paula D.; Fankhauser, Christian (2005): bHLH class transcription factors take centre stage in phytochrome signaling. In: *Trends in Plant Science* 10 (2), S. 51–54. DOI: 10.1016/j.tplants.2004.12.005.

Durand, Mickaël; Mainson, Dany; Porcheron, Benoît; Maurousset, Laurence; Lemoine, Rémi; Pourtau, Nathalie (2018): Carbon source-sink relationship in Arabidopsis thaliana: the role of sucrose transporters. In: *Planta* 247 (3), S. 587–611. DOI: 10.1007/s00425-017-2807-4.

Endo, Satoshi; Shinohara, Hidefumi; Matsubayashi, Yoshikatsu; Fukuda, Hiroo (2013): A novel pollen-pistil interaction conferring high-temperature tolerance during reproduction via CLE45 signaling. In: *Current Biology* 23 (17), S. 1670–1676. DOI: 10.1016/j.cub.2013.06.060.

Fang, Yitian; Chang, Jinke; Shi, Tao; Luo, Wenchun; Ou, Yang; Wan, Dongshi; Li, Jia (2021): Evolution of RGF/GLV/CLEL peptide hormones and their roles in land plant growth and regulation. In: *International Journal of Molecular Sciences* 22 (24). DOI: 10.3390/ijms222413372.

Fernandez, Ana; Drozdzecki, Andrzej; Hoogewijs, Kurt; Vassileva, Valya; Madder, Annemieke; Beeckman, Tom; Hilson, Pierre (2015): The GLV6/RGF8/CLEL2 peptide regulates early pericycle divisions during lateral root initiation. In: *Journal of Experimental Botany* 66 (17), S. 5245–5256. DOI: 10.1093/jxb/erv329.

References

Fernandez, Ana; Hilson, Pierre; Beeckman, Tom (2013): GOLVEN peptides as important regulatory signaling molecules of plant development. In: *Journal of Experimental Botany* 64 (17), S. 5263–5268. DOI: 10.1093/jxb/ert248.

Fernandez, Ana I.; Beeckman, Tom (2020): An MAP Kinase Cascade Downstream of RGF/GLV Peptides and Their RGI Receptors Regulates Root Development. In: *Molecular Plant* 13 (11), S. 1542–1544. DOI: 10.1016/j.molp.2020.10.009.

Fernandez, Ana I.; Vangheluwe, Nick; Xu, Ke; Jourquin, Joris; Claus, Lucas Alves Neubus; Morales-Herrera, Stefania et al. (2020): GOLVEN peptide signaling through RGI receptors and MPK6 restricts asymmetric cell division during lateral root initiation. In: *Nature Plants* 6 (5), S. 533–543. DOI: 10.1038/s41477-020-0645-z.

Ferrero, Lucía; Viola, Ivana L.; Ariel, Federico D.; Gonzalez, Daniel H. (2019): Class I TCP Transcription Factors Target the Gibberellin Biosynthesis Gene GA20ox1 and the Growth-Promoting Genes HBI1 and PRE6 during Thermomorphogenic Growth in Arabidopsis. In: *Plant & Cell Physiology* 60 (8), S. 1633–1645. DOI: 10.1093/pcp/pcz137.

Flohr, Aljoscha M.; Hackenbeck, Thomas; Schlueter, Claudia; Rogalla, Piere; Bullerdiek, Joern (2003): DNase I treatment of cDNA first strands prevents RT-PCR amplification of contaminating DNA sequences. In: *BioTechniques* 35 (5), 920-2, 924, 926. DOI: 10.2144/03355bm03.

Fuglsang, Anja T.; Kristensen, Astrid; Cuin, Tracey A.; Schulze, Waltraud X.; Persson, Jörgen; Thuesen, Kristina H. et al. (2014): Receptor kinase-mediated control of primary active proton pumping at the plasma membrane. In: *The Plant journal* 80 (6), S. 951–964. DOI: 10.1111/tpj.12680.

Fujita, Satoshi (2021): CASPARIAN STRIP INTEGRITY FACTOR (CIF) family peptides - regulator of plant extracellular barriers. In: *Peptides* 143, S. 170599. DOI: 10.1016/j.peptides.2021.170599.

Furumizu, Chihiro; Krabberød, Anders K.; Hammerstad, Marta; Alling, Renate M.; Wildhagen, Mari; Sawa, Shinichiro; Aalen, Reidunn B. (2021a): The sequenced genomes of nonflowering land plants reveal the innovative evolutionary history of peptide signaling. In: *The Plant Cell* 33 (9), S. 2915–2934. DOI: 10.1093/plcell/koab173.

Furumizu, Chihiro; Sawa, Shinichiro (2021b): The RGF/GLV/CLEL family of short peptides evolved through lineage-Specific losses and diversification and yet conserves its signaling role between vascular plants and bryophytes. In: *Frontiers in Plant Science* 12, S. 703012. DOI: 10.3389/fpls.2021.703012.

Geilfus, Christoph-Martin (2017): The pH of the Apoplast: Dynamic Factor with Functional Impact Under Stress. In: *Molecular Plant* 10 (11), S. 1371–1386. DOI: 10.1016/j.molp.2017.09.018.

Ghorbani, Sarieh; Hoogewijs, Kurt; Pečenková, Tamara; Fernandez, Ana; Inzé, Annelies; Eeckhout, Dominique et al. (2016): The SBT6.1 subtilase processes the GOLVEN1 peptide controlling cell elongation. In: *Journal of Experimental Botany* 67 (16), S. 4877–4887. DOI: 10.1093/jxb/erw241.

Giorgi, Julien de; Fuchs, Christelle; Iwasaki, Mayumi; Kim, Woohyun; Piskurewicz, Urszula; Gully, Kay et al. (2021): The Arabidopsis mature endosperm promotes seedling cuticle formation via release of sulfated peptides. In: *Developmental Cell* 56 (22), 3066-3081.e5. DOI: 10.1016/j.devcel.2021.10.005.

González-Carranza, Zinnia Haydee; Shahid, Ahmad Ali; Zhang, Li; Liu, Yang; Ninsuwan, Unchalee; Roberts, Jeremy Alan (2012): A novel approach to dissect the abscission process in Arabidopsis. In: *Plant Physiology* 160 (3), S. 1342–1356. DOI: 10.1104/pp.112.205955.

Grosse-Holz, Friederike; Kelly, Steven; Blaskowski, Svenja; Kaschani, Farnusch; Kaiser, Markus; van der Hoorn, Renier A. L. (2018): The transcriptome, extracellular proteome and active secretome of agroinfiltrated Nicotiana benthamiana uncover a large, diverse protease repertoire. In: *Plant Biotechnology Journal* 16 (5), S. 1068–1084. DOI: 10.1111/pbi.12852.

Han, Huibin; Glazunova, Alina; Wang, Guodong (2023): pH regulates peptide-receptor perception. In: *Trends in plant science*. DOI: 10.1016/j.tplants.2023.04.015.

Hanai, H.; Matsuno, T.; Yamamoto, M.; Matsubayashi, Y.; Kobayashi, T.; Kamada, H.; Sakagami, Y. (2000): A secreted peptide growth factor, phytosulfokine, acting as a stimulatory factor of carrot somatic embryo formation. In: *Plant & Cell Physiology* 41 (1), S. 27–32. DOI: 10.1093/pcp/41.1.27.

Hander, Tim; Fernández-Fernández, Álvaro D.; Kumpf, Robert P.; Willems, Patrick; Schatowitz, Hendrik; Rombaut, Debbie et al. (2019): Damage on plants activates Ca2+-dependent metacaspases for release of immunomodulatory peptides. In: *Science* 363 (6433). DOI: 10.1126/science.aar7486.

Hao, Zhaodong; Wu, Hua; Zheng, Renhua; Li, Rui; Zhu, Zeli; Chen, Ya et al. (2023): The plant peptide hormone phytosulfokine promotes somatic embryogenesis by maintaining redox homeostasis in Cunninghamia lanceolata. In: *The Plant journal* 113 (4), S. 716–733. DOI: 10.1111/tpj.16077.

Hartmann, Jens; Stührwohldt, Nils; Dahlke, Renate I.; Sauter, Margret (2013): Phytosulfokine control of growth occurs in the epidermis, is likely to be non-cell autonomous and is dependent on brassinosteroids. In: *The Plant journal* 73 (4), S. 579–590. DOI: 10.1111/tpj.12056.

Hazman, Mohamed; Hause, Bettina; Eiche, Elisabeth; Riemann, Michael; Nick, Peter (2016): Different forms of osmotic stress evoke qualitatively different responses in rice. In: *Journal of Plant Physiology* 202, S. 45–56. DOI: 10.1016/j.jplph.2016.05.027.

He, Z.; Wang, Z. Y.; Li, J.; Zhu, Q.; Lamb, C.; Ronald, P.; Chory, J. (2000): Perception of brassinosteroids by the extracellular domain of the receptor kinase BRI1. In: *Science (New York, N.Y.)* 288 (5475), S. 2360–2363. DOI: 10.1126/science.288.5475.2360.

Hieta, Reija; Myllyharju, Johanna (2002): Cloning and characterization of a low molecular weight prolyl 4-hydroxylase from Arabidopsis thaliana. Effective hydroxylation of proline-rich, collagen-like, and hypoxia-inducible transcription factor alpha-like peptides. In: *The Journal of Biological Chemistry* 277 (26), S. 23965–23971. DOI: 10.1074/jbc.M201865200.

Hohmann, Ulrich; Lau, Kelvin; Hothorn, Michael (2017): The Structural Basis of Ligand Perception and Signal Activation by Receptor Kinases. In: *Annual Review of Plant Biology* 68, S. 109–137. DOI: 10.1146/annurev-arplant-042916-040957.

Holton, T. A.; Cornish, E. C. (1995): Genetics and Biochemistry of Anthocyanin Biosynthesis. In: *The Plant Cell* 7 (7), S. 1071–1083. DOI: 10.1105/tpc.7.7.1071.

Holzwart, Eleonore; Huerta, Apolonio Ignacio; Glöckner, Nina; Garnelo Gómez, Borja; Wanke, Friederike; Augustin, Sebastian et al. (2018): BRI1 controls vascular cell fate in the Arabidopsis root through RLP44 and phytosulfokine signaling. In: *Proceedings of the National Academy of Sciences of the United States of America* 115 (46), S. 11838–11843. DOI: 10.1073/pnas.1814434115.

Hou, Shuguo; Liu, Derui; He, Ping (2021): Phytocytokines function as immunological modulators of plant immunity. In: *Stress biology* 1 (1), S. 8. DOI: 10.1007/s44154-021-00009-y.

Hou, Shuguo; Wang, Xin; Chen, Donghua; Yang, Xue; Wang, Mei; Turrà, David et al. (2014): The secreted peptide PIP1 amplifies immunity through receptor-like kinase 7. In: *PLoS pathogens* 10 (9), e1004331. DOI: 10.1371/journal.ppat.1004331.

Huffaker, Alisa; Ryan, Clarence A. (2007): Endogenous peptide defense signals in Arabidopsis differentially amplify signaling for the innate immune response. In: *Proceedings of the National Academy of Sciences of the United States of America* 104 (25), S. 10732–10736. DOI: 10.1073/pnas.0703343104.

Huq, Enamul; Al-Sady, Bassem; Hudson, Matthew; Kim, Chanhong; Apel, Klaus; Quail, Peter H. (2004): Phytochrome-interacting factor 1 is a critical bHLH regulator of chlorophyll biosynthesis. In: *Science* 305 (5692), S. 1937–1941. DOI: 10.1126/science.1099728.

Hutin, Claire; Nussaume, Laurent; Moise, Nicolae; Moya, Ismaël; Kloppstech, Klaus; Havaux, Michel (2003): Early light-induced proteins protect Arabidopsis from photooxidative stress. In: *Proceedings of the National Academy of Sciences of the United States of America* 100 (8), S. 4921–4926. DOI: 10.1073/pnas.0736939100.

Huttner, W. B. (1982): Sulphation of tyrosine residues-a widespread modification of proteins. In: *Nature* 299 (5880), S. 273–276. DOI: 10.1038/299273a0.

Ibañez, Carla; Delker, Carolin; Martinez, Cristina; Bürstenbinder, Katharina; Janitza, Philipp; Lippmann, Rebecca et al. (2018): Brassinosteroids Dominate Hormonal Regulation of Plant Thermomorphogenesis via BZR1. In: *Current Biology* 28 (2), 303-310.e3. DOI: 10.1016/j.cub.2017.11.077.

Igarashi, Daisuke; Tsuda, Kenichi; Katagiri, Fumiaki (2012): The peptide growth factor, phytosulfokine, attenuates pattern-triggered immunity. In: *The Plant journal* 71 (2), S. 194–204. DOI: 10.1111/j.1365-313X.2012.04950.x.

References

Jang, In-Cheol; Yang, Jun-Yi; Seo, Hak Soo; Chua, Nam-Hai (2005): HFR1 is targeted by COP1 E3 ligase for post-translational proteolysis during phytochrome A signaling. In: *Genes & Development* 19 (5), S. 593–602. DOI: 10.1101/gad.1247205.

Jeon, Byeong Wook; Kim, Jin Sun; Oh, Eunkyoo; Kang, Na Young; Kim, Jungmook (2023): ROOT MERISTEM GROWTH FACTOR1 (RGF1)-RGF1 INSENSITIVE 1 peptide-receptor pair inhibits lateral root development via the MPK6-PUCHI module in Arabidopsis. In: *Journal of Experimental Botany* 74 (5), S. 1475–1488. DOI: 10.1093/jxb/erac495.

Jeon, Byeong Wook; Kim, Min-Jung; Pandey, Shashank K.; Oh, Eunkyoo; Seo, Pil Joon; Kim, Jungmook (2021): Recent advances in peptide signaling during Arabidopsis root development. In: *Journal of Experimental Botany* 72 (8), S. 2889–2902. DOI: 10.1093/jxb/erab050.

Jiao, Yuling; Lau, On Sun; Deng, Xing Wang (2007): Light-regulated transcriptional networks in higher plants. In: *Nature reviews. Genetics* 8 (3), S. 217–230. DOI: 10.1038/nrg2049.

Jung, Jae-Hoon; Domijan, Mirela; Klose, Cornelia; Biswas, Surojit; Ezer, Daphne; Gao, Mingjun et al. (2016): Phytochromes function as thermosensors in Arabidopsis. In: *Science* 354 (6314), S. 886–889. DOI: 10.1126/science.aaf6005.

Kang, Jun; Yu, Haopeng; Tian, Caihuan; Zhou, Wenkun; Li, Chuanyou; Jiao, Yuling; Liu, Dong (2014): Suppression of Photosynthetic Gene Expression in Roots Is Required for Sustained Root Growth under Phosphate Deficiency. In: *Plant Physiology* 165 (3), S. 1156–1170. DOI: 10.1104/pp.114.238725.

Kaufmann, Christine; Sauter, Margret (2017): Pull-down Assay to Characterize Ca2+/Calmodulin Binding to Plant Receptor Kinases: Springer New York (Methods in Molecular Biology, 1621). In: *Methods in molecular biology (Clifton, N.J.)*, S. 151–159.

Kaufmann, Christine; Sauter, Margret (2019): Sulfated plant peptide hormones. In: *Journal of Experimental Botany* 70 (16), S. 4267–4277. DOI: 10.1093/jxb/erz292.

Kaufmann, Christine; Stührwohldt, Nils; Sauter, Margret (2021): Tyrosylprotein sulfotransferase-dependent and -independent regulation of root development and signaling by PSK LRR receptor kinases in Arabidopsis. In: *Journal of Experimental Botany* 72 (15), S. 5508–5521. DOI: 10.1093/jxb/erab233.

Kim, Jin Sun; Jeon, Byeong Wook; Kim, Jungmook (2021): Signaling Peptides Regulating Abiotic Stress Responses in Plants. In: *Frontiers in Plant Science* 12, S. 704490. DOI: 10.3389/fpls.2021.704490.

Kivirikko, K. I.; Myllylä, R.; Pihlajaniemi, T. (1989): Protein hydroxylation: prolyl 4-hydroxylase, an enzyme with four cosubstrates and a multifunctional subunit. In: *FASEB journal: official publication of the Federation of American Societies for Experimental Biology* 3 (5), S. 1609–1617.

Kobayashi, Koichi; Sasaki, Daichi; Noguchi, Ko; Fujinuma, Daiki; Komatsu, Hirohisa; Kobayashi, Masami et al. (2013): Photosynthesis of root chloroplasts developed in Arabidopsis lines overexpressing GOLDEN2-LIKE transcription factors. In: *Plant & Cell Physiology* 54 (8), S. 1365–1377. DOI: 10.1093/pcp/pct086.

Koini, Maria A.; Alvey, Liz; Allen, Trudie; Tilley, Ceinwen A.; Harberd, Nicholas P.; Whitelam, Garry C.; Franklin, Keara A. (2009): High temperature-mediated adaptations in plant architecture require the bHLH transcription factor PIF4. In: *Current Biology* 19 (5), S. 408–413. DOI: 10.1016/j.cub.2009.01.046.

Komori, Ryota; Amano, Yukari; Ogawa-Ohnishi, Mari; Matsubayashi, Yoshikatsu (2009): Identification of tyrosylprotein sulfotransferase in Arabidopsis. In: *Proceedings of the National Academy of Sciences of the United States of America* 106 (35), S. 15067–15072. DOI: 10.1073/pnas.0902801106.

Koussevitzky, Shai; Nott, Ajit; Mockler, Todd C.; Hong, Fangxin; Sachetto-Martins, Gilberto; Surpin, Marci et al. (2007): Signals from chloroplasts converge to regulate nuclear gene expression. In: *Science (New York, N.Y.)* 316 (5825), S. 715–719.

Krawczyk, Hannah Elisa (2021): Identification and Characterisation of Lipid Droplet-Localised Proteins.

Kutschmar, Anke; Rzewuski, Guillaume; Stührwohldt, Nils; Beemster, Gerrit T. S.; Inzé, Dirk; Sauter, Margret (2009): PSK-α promotes root growth in Arabidopsis. In: *New Phytologist* 181 (4), S. 820–831. DOI: 10.1111/j.1469-8137.2008.02710.x.

Ladwig, Friederike; Dahlke, Renate I.; Stührwohldt, Nils; Hartmann, Jens; Harter, Klaus; Sauter, Margret (2015): Phytosulfokine Regulates Growth in Arabidopsis through a Response Module at the Plasma Membrane That Includes CYCLIC NUCLEOTIDE-GATED CHANNEL17, H+-ATPase, and BAK1. In: *The Plant Cell* 27 (6), S. 1718–1729. DOI: 10.1105/tpc.15.00306.

Lantzouni, Ourania; Alkofer, Angela; Falter-Braun, Pascal; Schwechheimer, Claus (2020): GROWTH-REGULATING FACTORS Interact with DELLAs and Regulate Growth in Cold Stress. In: *The Plant Cell* 32 (4), S. 1018–1034. DOI: 10.1105/tpc.19.00784.

Lee, Jungeun; He, Kun; Stolc, Viktor; Lee, Horim; Figueroa, Pablo; Gao, Ying et al. (2007): Analysis of transcription factor HY5 genomic binding sites revealed its hierarchical role in light regulation of development. In: *The Plant Cell* 19 (3), S. 731–749. DOI: 10.1105/tpc.106.047688.

Lee, Sanghwa; Wang, Wenli; Huq, Enamul (2021): Spatial regulation of thermomorphogenesis by HY5 and PIF4 in Arabidopsis. In: *Nature communications* 12 (1), S. 3656. DOI: 10.1038/s41467-021-24018-7.

Legris, Martina; Klose, Cornelia; Burgie, E. Sethe; Rojas, Cecilia Costigliolo Rojas; Neme, Maximiliano; Hiltbrunner, Andreas et al. (2016): Phytochrome B integrates light and temperature signals in Arabidopsis. In: *Science* 354 (6314), S. 897–900. DOI: 10.1126/science.aaf5656.

Leivar, Pablo; Monte, Elena; Oka, Yoshito; Liu, Tiffany; Carle, Christine; Castillon, Alicia et al. (2008): Multiple phytochrome-interacting bHLH transcription factors repress premature seedling photomorphogenesis in darkness. In: *Current Biology* 18 (23), S. 1815–1823. DOI: 10.1016/j.cub.2008.10.058.

Leivar, Pablo; Tepperman, James M.; Monte, Elena; Calderon, Robert H.; Liu, Tiffany L.; Quail, Peter H. (2009): Definition of early transcriptional circuitry involved in light-induced reversal of PIF-imposed repression of photomorphogenesis in young Arabidopsis seedlings. In: *The Plant Cell* 21 (11), S. 3535–3553. DOI: 10.1105/tpc.109.070672.

Li, Kunlun; Yu, Renbo; Fan, Liu-Min; Wei, Ning; Chen, Haodong; Deng, Xing Wang (2016): DELLA-mediated PIF degradation contributes to coordination of light and gibberellin signaling in Arabidopsis. In: *Nature communications* 7, S. 11868. DOI: 10.1038/ncomms11868.

Li, Xu; Liang, Tong; Liu, Hongtao (2022): How plants coordinate their development in response to light and temperature signals. In: *The Plant Cell* 34 (3), S. 955–966. DOI: 10.1093/plcell/koab302.

Liu, Li; Song, Wen; Huang, Shijia; Jiang, Kai; Moriwaki, Yoshitaka; Wang, Yichuan et al. (2022): Extracellular pH sensing by plant cell-surface peptide-receptor complexes. In: *Cell* 185 (18), 3341-3355.e13. DOI: 10.1016/j.cell.2022.07.012.

Liu, Xiaoqin; Liu, Renlu; Li, Yue; Shen, Xing; Zhong, Shangwei; Shi, Hui (2017): EIN3 and PIF3 form an interdependent module that represses chloroplast development in buried seedlings. In: *The Plant Cell* 29 (12), S. 3051–3067. DOI: 10.1105/tpc.17.00508.

Liu, Yumin; Zhang, Danping; Li, Mei; Yan, Junhui; Luo, Li; Yu, Liangliang (2019): Overexpression of PSK-γ in Arabidopsis promotes growth without influencing pattern-triggered immunity. In: *Plant Signaling & Behavior* 14 (12), S. 1684423. DOI: 10.1080/15592324.2019.1684423.

Livak, K. J.; Schmittgen, T. D. (2001): Analysis of relative gene expression data using real-time quantitative PCR and the 2(-Delta Delta C(T)) Method. In: *Methods (San Diego, Calif.)* 25 (4), S. 402–408. DOI: 10.1006/meth.2001.1262.

Loivamäki (2010): A role for PSK signaling in wounding and microbial interactions in Arabidopsis. In: *Physiologia Plantarum* 139, S. 348.

Loreti, Elena; Povero, Giovanni; Novi, Giacomo; Solfanelli, Cinzia; Alpi, Amedeo; Perata, Pierdomenico (2008): Gibberellins, jasmonate and abscisic acid modulate the sucrose-induced expression of anthocyanin biosynthetic genes in Arabidopsis. In: *New Phytologist* 179 (4), S. 1004–1016. DOI: 10.1111/j.1469-8137.2008.02511.x.

Lorrain, Séverine; Allen, Trudie; Duek, Paula D.; Whitelam, Garry C.; Fankhauser, Christian (2008): Phytochrome-mediated inhibition of shade avoidance involves degradation of growth-promoting bHLH transcription factors. In: *The Plant Journal* 53 (2), S. 312–323. DOI: 10.1111/j.1365-313X.2007.03341.x.

Lu, Xiaoting; Shi, Hongyong; Ou, Yang; Cui, Yanwei; Chang, Jinke; Peng, Liang et al. (2020): RGF1-RGI1, a Peptide-Receptor Complex, Regulates Arabidopsis Root Meristem Development via a MAPK Signaling Cascade. In: *Molecular Plant* 13 (11), S. 1594–1607. DOI: 10.1016/j.molp.2020.09.005.

Mahon, P.; Bateman, A. (2000): The PA domain: a protease-associated domain. In: *Protein Science* 9 (10), S. 1930–1934. DOI: 10.1110/ps.9.10.1930.

Maier, Alexander; Schrader, Andrea; Kokkelink, Leonie; Falke, Christian; Welter, Bastian; Iniesto, Elisa et al. (2013): Light and the E3 ubiquitin ligase COP1/SPA control the protein stability of the MYB transcription factors PAP1 and PAP2 involved in anthocyanin accumulation in Arabidopsis. In: *The Plant Journal* 74 (4), S. 638–651. DOI: 10.1111/tpj.12153.

Martín, Guiomar; Leivar, Pablo; Ludevid, Dolores; Tepperman, James M.; Quail, Peter H.; Monte, Elena (2016): Phytochrome and retrograde signaling pathways converge to antagonistically regulate a light-induced transcriptional network. In: *Nature communications* 7, S. 11431. DOI: 10.1038/ncomms11431.

Matsubayashi, Y.; Sakagami, Y. (1996): Phytosulfokine, sulfated peptides that induce the proliferation of single mesophyll cells of Asparagus officinalis L. In: *Proceedings of the National Academy of Sciences of the United States of America* 93 (15), S. 7623–7627. DOI: 10.1073/pnas.93.15.7623.

Matsubayashi, Y.; Takagi, L.; Omura, N.; Morita, A.; Sakagami, Y. (1999): The endogenous sulfated pentapeptide phytosulfokine-alpha stimulates tracheary element differentiation of isolated mesophyll cells of zinnia. In: *Plant Physiology* 120 (4), S. 1043–1048. DOI: 10.1104/pp.120.4.1043.

Matsubayashi, Yoshikatsu (2011): Small post-translationally modified Peptide signals in Arabidopsis. In: *The arabidopsis book* 9, e0150. DOI: 10.1199/tab.0150.

Matsubayashi, Yoshikatsu (2014): Posttranslationally modified small-peptide signals in plants. In: *Annual Review of Plant Biology* 65 (1), S. 385–413. DOI: 10.1146/annurev-arplant-050312-120122.

Matsubayashi, Yoshikatsu; Ogawa, Mari; Kihara, Hitomi; Niwa, Masaaki; Sakagami, Youji (2006): Disruption and overexpression of Arabidopsis phytosulfokine receptor gene affects cellular longevity and potential for growth. In: *Plant Physiology* 142 (1), S. 45–53. DOI: 10.1104/pp.106.081109.

Matsuzaki, Yo; Ogawa-Ohnishi, Mari; Mori, Ayaka; Matsubayashi, Yoshikatsu (2010): Secreted peptide signals required for maintenance of root stem cell niche in Arabidopsis. In: *Science* 329 (5995), S. 1065–1067. DOI: 10.1126/science.1191132.

Meng, Ling; Buchanan, Bob B.; Feldman, Lewis J.; Luan, Sheng (2012): CLE-like (CLEL) peptides control the pattern of root growth and lateral root development in Arabidopsis. In: *Proceedings of the National Academy of Sciences of the United States of America* 109 (5), S. 1760–1765. DOI: 10.1073/pnas.1119864109.

Meskauskiene, R.; Nater, M.; Goslings, D.; Kessler, F.; op den Camp, R.; Apel, K. (2001): FLU: a negative regulator of chlorophyll biosynthesis in Arabidopsis thaliana. In: *Proceedings of the National Academy of Sciences of the United States of America* 98 (22), S. 12826–12831. DOI: 10.1073/pnas.221252798.

Meyer, Michael; Leptihn, Sebastian; Welz, Max; Schaller, Andreas (2016): Functional Characterization of Propeptides in Plant Subtilases as Intramolecular Chaperones and Inhibitors of the Mature Protease. In: *The Journal of Biological Chemistry* 291 (37), S. 19449–19461. DOI: 10.1074/jbc.M116.744151.

Monte, Elena; Al-Sady, Bassem; Leivar, Pablo; Quail, Peter H. (2007): Out of the dark: how the PIFs are unmasking a dual temporal mechanism of phytochrome signaling. In: *Journal of Experimental Botany* 58 (12), S. 3125–3133. DOI: 10.1093/jxb/erm186.

Moore, Kevin L. (2003): The biology and enzymology of protein tyrosine O-sulfation. In: *The Journal of Biological Chemistry* 278 (27), S. 24243–24246. DOI: 10.1074/jbc.R300008200.

Moore, Kevin L. (2009): Protein tyrosine sulfation: a critical posttranslation modification in plants and animals. In: *Proceedings of the National Academy of Sciences of the United States of America* 106 (35), S. 14741–14742. DOI: 10.1073/pnas.0908376106.

Mosher, Stephen; Seybold, Heike; Rodriguez, Patricia; Stahl, Mark; Davies, Kelli A.; Dayaratne, Sajeewani et al. (2013): The tyrosine-sulfated peptide receptors PSKR1 and PSY1R modify the immunity of Arabidopsis to biotrophic and necrotrophic pathogens in an antagonistic manner. In: *The Plant journal* 73 (3), S. 469–482. DOI: 10.1111/tpj.12050.

Motose, Hiroyasu; Iwamoto, Kuninori; Endo, Satoshi; Demura, Taku; Sakagami, Youji; Matsubayashi, Yoshikatsu et al. (2009): Involvement of phytosulfokine in the attenuation of stress response during the transdifferentiation of zinnia mesophyll cells into tracheary elements. In: *Plant Physiology* 150 (1), S. 437–447. DOI: 10.1104/pp.109.135954.

Mullet, J. E. (1988): Chloroplast development and gene expression. In: *Annual Review of Plant Physiology and Plant Molecular Biology* 39 (1), S. 475–502. DOI: 10.1146/annurev.pp.39.060188.002355.

Murashige, Toshio; Skoog, Folke (1962): A Revised Medium for Rapid Growth and Bio Assays with Tobacco Tissue Cultures. In: *Physiologia Plantarum* 15 (3), S. 473–497. DOI: 10.1111/j.1399-3054.1962.tb08052.x.

Nakabayashi, Ryo; Yonekura-Sakakibara, Keiko; Urano, Kaoru; Suzuki, Makoto; Yamada, Yutaka; Nishizawa, Tomoko et al. (2014): Enhancement of oxidative and drought tolerance in Arabidopsis by overaccumulation of antioxidant flavonoids. In: *The Plant journal* 77 (3), S. 367–379. DOI: 10.1111/tpj.12388.

Nakaminami, Kentaro; Okamoto, Masanori; Higuchi-Takeuchi, Mieko; Yoshizumi, Takeshi; Yamaguchi, Yube; Fukao, Yoichiro et al. (2018): AtPep3 is a hormone-like peptide that plays a role in the salinity stress tolerance of plants. In: *Proceedings of the National Academy of Sciences of the United States of America* 115 (22), S. 5810–5815. DOI: 10.1073/pnas.1719491115.

Nakaminami, Kentaro; Okamoto, Masanori; Higuchi-Takeuchi, Mieko; Yoshizumi, Takeshi; Yamaguchi, Yube; Fukao, Yoichiro et al. (2018): AtPep3 is a hormone-like peptide that plays a role in the salinity stress tolerance of plants. In: *Proceedings of the National Academy of Sciences of the United States of America* 115 (22), S. 5810–5815. DOI: 10.1073/pnas.1719491115.

Nakata, Masaru; Mitsuda, Nobutaka; Herde, Marco; Koo, Abraham J. K.; Moreno, Javier E.; Suzuki, Kaoru et al. (2013): A bHLH-type transcription factor, ABA-INDUCIBLE BHLH-TYPE TRANSCRIPTION FACTOR/JA-ASSOCIATED MYC2-LIKE1, acts as a repressor to negatively regulate jasmonate signaling in arabidopsis. In: *The Plant Cell* 25 (5), S. 1641–1656. DOI: 10.1105/tpc.113.111112.

Nakayama, Takuya; Shinohara, Hidefumi; Tanaka, Mina; Baba, Koki; Ogawa-Ohnishi, Mari; Matsubayashi, Yoshikatsu (2017): A peptide hormone required for Casparian strip diffusion barrier formation in Arabidopsis roots. In: *Science* 355 (6322), S. 284–286. DOI: 10.1126/science.aai9057.

Nusinow, Dmitri A.; Helfer, Anne; Hamilton, Elizabeth E.; King, Jasmine J.; Imaizumi, Takato; Schultz, Thomas F. et al. (2011): The ELF4-ELF3-LUX complex links the circadian clock to diurnal control of hypocotyl growth. In: *Nature* 475 (7356), S. 398–402. DOI: 10.1038/nature10182.

Ogawa-Ohnishi, Mari; Matsushita, Wataru; Matsubayashi, Yoshikatsu (2013): Identification of three hydroxyproline O-arabinosyltransferases in Arabidopsis thaliana. In: *Nature chemical biology* 9 (11), S. 726–730. DOI: 10.1038/nchembio.1351.

Ogawa-Ohnishi, Mari; Yamashita, Tomohide; Kakita, Mitsuru; Nakayama, Takuya; Ohkubo, Yuri; Hayashi, Yoko et al. (2022): Peptide ligand-mediated trade-off between plant growth and stress response. In: *Science* 378 (6616), S. 175–180. DOI: 10.1126/science.abq5735.

Oh, Eunkyoo; Kim, Jonghyun; Park, Eunae; Kim, Jeong-Il; Kang, Changwon; Choi, Giltsu (2004): PIL5, a phytochrome-interacting basic helix-loop-helix protein, is a key negative regulator of seed germination in Arabidopsis thaliana. In: *The Plant Cell* 16 (11), S. 3045–3058. DOI: 10.1105/tpc.104.025163.

Oh, Eunkyoo; Zhu, Jia-Ying; Wang, Zhi-Yong (2012): Interaction between BZR1 and PIF4 integrates brassinosteroid and environmental responses. In: *Nature cell biology* 14 (8), S. 802–809. DOI: 10.1038/ncb2545.

Ohkubo, Yuri; Tanaka, Mina; Tabata, Ryo; Ogawa-Ohnishi, Mari; Matsubayashi, Yoshikatsu (2017): Shoot-to-root mobile polypeptides involved in systemic regulation of nitrogen acquisition. In: *Nature Plants* 3, S. 17029. DOI: 10.1038/nplants.2017.29.

Ohyama, Kentaro; Ogawa, Mari; Matsubayashi, Yoshikatsu (2008): Identification of a biologically active, small, secreted peptide in Arabidopsis by in silico gene screening, followed by LC-MS-based structure analysis. In: *The Plant journal* 55 (1), S. 152–160. DOI: 10.1111/j.1365-313X.2008.03464.x.

Okuda, Satohiro; Fujita, Satoshi; Moretti, Andrea; Hohmann, Ulrich; Doblas, Verónica G.; Ma, Yan et al. (2020): Molecular mechanism for the recognition of sequence-divergent CIF peptides by the plant receptor kinases GSO1/SGN3 and GSO2. In: *Proceedings of the National Academy of Sciences of the United States of America* 117 (5), S. 2693–2703. DOI: 10.1073/pnas.1911553117.

References

Orosa-Puente, Beatriz; Leftley, Nicola; Wangenheim, Daniel von; Banda, Jason; Srivastava, Anjil K.; Hill, Kristine et al. (2018): Root branching toward water involves posttranslational modification of transcription factor ARF7. In: *Science* 362 (6421), S. 1407–1410. DOI: 10.1126/science.aau3956.

Osterlund, M. T.; Hardtke, C. S.; Wei, N.; Deng, X. W. (2000): Targeted destabilization of HY5 during light-regulated development of Arabidopsis. In: *Nature* 405 (6785), S. 462–466. DOI: 10.1038/35013076.

Ou, Yang; Lu, Xiaoting; Zi, Quaner; Xun, Qingqing; Zhang, Jingjie; Wu, Yujun et al. (2016): RGF1 INSENSITIVE 1 to 5, a group of LRR receptor-like kinases, are essential for the perception of root meristem growth factor 1 in Arabidopsis thaliana. In: *Cell Research* 26 (6), S. 686–698. DOI: 10.1038/cr.2016.63.

Ou, Yang; Tao, Bingqing; Wu, Yujun; Cai, Zeping; Li, Huiqiang; Li, Meizhen et al. (2022): Essential roles of SERKs in the ROOT MERISTEM GROWTH FACTOR-mediated signaling pathway. In: *Plant Physiology* 189 (1), S. 165–177. DOI: 10.1093/plphys/kiac036.

Patharkar, O. Rahul; Walker, John C. (2016): Core mechanisms regulating developmentally timed and environmentally triggered abscission. In: *Plant Physiology* 172 (1), S. 510–520. DOI: 10.1104/pp.16.01004.

Pearce, G.; Strydom, D.; Johnson, S.; Ryan, C. A. (1991): A polypeptide from tomato leaves induces wound-inducible proteinase inhibitor proteins. In: *Science (New York, N.Y.)* 253 (5022), S. 895–897. DOI: 10.1126/science.253.5022.895.

Pervaiz, Tariq; Songtao, Jiu; Faghihi, Faezeh; Haider, Muhammad Salman; Fang, Jinggui (2017): Naturally Occurring Anthocyanin, Structure, Functions and Biosynthetic Pathway in Fruit Plants. In: *J Plant Biochem Physiol* 05 (02). DOI: 10.4172/2329-9029.1000187.

Pfaffl, M. W. (2001): A new mathematical model for relative quantification in real-time RT-PCR. In: *Nucleic Acids Research* 29 (9), e45. DOI: 10.1093/nar/29.9.e45.

Pham, Vinh Ngoc; Kathare, Praveen Kumar; Huq, Enamul (2018): Dynamic regulation of PIF5 by COP1-SPA complex to optimize photomorphogenesis in Arabidopsis. In: *The Plant journal* 96 (2), S. 260–273. DOI: 10.1111/tpj.14074.

Pham, Vinh Ngoc; Kathare, Praveen Kumar; Huq, Enamul (2018): Phytochromes and Phytochrome Interacting Factors. In: *Plant Physiology* 176 (2), S. 1025–1038. DOI: 10.1104/pp.17.01384.

Pruitt, Rory N.; Joe, Anna; Zhang, Weiguo; Feng, Wei; Stewart, Valley; Schwessinger, Benjamin et al. (2017): A microbially derived tyrosine-sulfated peptide mimics a plant peptide hormone. In: *New Phytologist* 215 (2), S. 725–736. DOI: 10.1111/nph.14609.

Quint, Marcel; Delker, Carolin; Franklin, Keara A.; Wigge, Philip A.; Halliday, Karen J.; van Zanten, Martijn (2016): Molecular and genetic control of plant thermomorphogenesis. In: *Nature Plants* 2, S. 15190. DOI: 10.1038/nplants.2015.190.

Rautengarten, Carsten; Steinhauser, Dirk; Büssis, Dirk; Stintzi, Annick; Schaller, Andreas; Kopka, Joachim; Altmann, Thomas (2005): Inferring hypotheses on functional relationships of genes: Analysis of the Arabidopsis thaliana subtilase gene family. In: *PLoS Computational Biology* 1 (4), e40. DOI: 10.1371/journal.pcbi.0010040.

Reichardt, S.; Piepho, H-P; Stintzi, A.; Schaller, A. (2020): Peptide signaling for drought-induced tomato flower drop. In: *Science* 367 (6485), S. 1482–1485. DOI: 10.1126/science.aaz5641.

Richter, Andreas S.; Tohge, Takayuki; Fernie, Alisdair R.; Grimm, Bernhard (2020): The genomes uncoupled-dependent signaling pathway coordinates plastid biogenesis with the synthesis of anthocyanins. In: *Philosophical transactions of the Royal Society of London. Series B, Biological sciences* 375 (1801), S. 20190403. DOI: 10.1098/rstb.2019.0403.

Richter, René; Behringer, Carina; Müller, Isabel Karin; Schwechheimer, Claus (2010): The GATA-type transcription factors GNC and GNL/CGA1 repress gibberellin signaling downstream from DELLA proteins and PHYTOCHROME-INTERACTING FACTORS. In: *Genes & Development* 24 (18), S. 2093–2104. DOI: 10.1101/gad.594910.

Rodiuc, Natalia; Barlet, Xavier; Hok, Sophie; Perfus-Barbeoch, Laetitia; Allasia, Valérie; Engler, Gilbert et al. (2016): Evolutionarily distant pathogens require the

Arabidopsis phytosulfokine signaling pathway to establish disease. In: *Plant, Cell & Environment* 39 (7), S. 1396–1407. DOI: 10.1111/pce.12627.

Rojo, Enrique; Sharma, Vijay K.; Kovaleva, Valentina; Raikhel, Natasha V.; Fletcher, Jennifer C. (2002): CLV3 is localized to the extracellular space, where it activates the Arabidopsis CLAVATA stem cell signaling pathway. In: *The Plant Cell* 14 (5), S. 969–977. DOI: 10.1105/tpc.002196.

Rotsch, Alexander H.; Kopka, Joachim; Feussner, Ivo; Ischebeck, Till (2017): Central metabolite and sterol profiling divides tobacco male gametophyte development and pollen tube growth into eight metabolic phases. In: *The Plant journal* 92 (1), S. 129–146. DOI: 10.1111/tpj.13633.

Royek, Stefanie; Bayer, Martin; Pfannstiel, Jens; Pleiss, Jürgen; Ingram, Gwyneth; Stintzi, Annick; Schaller, Andreas (2022): Processing of a plant peptide hormone precursor facilitated by posttranslational tyrosine sulfation. In: *Proceedings of the National Academy of Sciences of the United States of America* 119 (16), e2201195119. DOI: 10.1073/pnas.2201195119.

Sakamoto, Wataru; Miyagishima, Shin-Ya; Jarvis, Paul (2008): Chloroplast biogenesis: control of plastid development, protein import, division and inheritance. In: *The arabidopsis book* 6, e0110. DOI: 10.1199/tab.0110.

Sakuraba, Yasuhito; Jeong, Jinkil; Kang, Min-Young; Kim, Junghyun; Paek, Nam-Chon; Choi, Giltsu (2014): Phytochrome-interacting transcription factors PIF4 and PIF5 induce leaf senescence in Arabidopsis. In: *Nature communications* 5, S. 4636. DOI: 10.1038/ncomms5636.

Sameer, Aga Syed; Nissar, Saniya (2021): Toll-Like Receptors (TLRs): Structure, functions, signaling, and role of their polymorphisms in colorectal cancer susceptibility. In: *BioMed research international* 2021, S. 1157023. DOI: 10.1155/2021/1157023.

Santiago, Julia; Brandt, Benjamin; Wildhagen, Mari; Hohmann, Ulrich; Hothorn, Ludwig A.; Butenko, Melinka A.; Hothorn, Michael (2016): Mechanistic insight into a peptide hormone signaling complex mediating floral organ abscission. In: *eLife* 5. DOI: 10.7554/eLife.15075.

Santiago, Julia; Henzler, Christine; Hothorn, Michael (2013): Molecular mechanism for plant steroid receptor activation by somatic embryogenesis co-receptor kinases. In: *Science* 341 (6148), S. 889–892. DOI: 10.1126/science.1242468.

Schägger, Hermann (2006): Tricine-SDS-PAGE. In: *Nature Protocols* 1 (1), S. 16–22. DOI: 10.1038/nprot.2006.4.

Schaller, Andreas; Stintzi, Annick; Rivas, Susana; Serrano, Irene; Chichkova, Nina V.; Vartapetian, Andrey B. et al. (2018): From structure to function - a family portrait of plant subtilases. In: *New Phytologist* 218 (3), S. 901–915. DOI: 10.1111/nph.14582.

Schardon, Katharina; Hohl, Mathias; Graff, Lucile; Pfannstiel, Jens; Schulze, Waltraud; Stintzi, Annick; Schaller, Andreas (2016): Precursor processing for plant peptide hormone maturation by subtilisin-like serine proteinases. In: *Science* 354 (6319), S. 1594–1597. DOI: 10.1126/science.aai8550.

Sham, Arjun; Moustafa, Khaled; Al-Ameri, Salma; Al-Azzawi, Ahmed; Iratni, Rabah; AbuQamar, Synan (2015): Identification of Arabidopsis candidate genes in response to biotic and abiotic stresses using comparative microarrays. In: *PLOS ONE* 10 (5), e0125666. DOI: 10.1371/journal.pone.0125666.

Shao, Ling; Shu, Zhan; Sun, Shu-Lan; Peng, Chang-Lian; Wang, Xiao-Jing; Lin, Zhi-Fang (2007): Antioxidation of Anthocyanins in Photosynthesis Under High Temperature Stress. In: *Journal of Integrative Plant Biology* 49 (9), S. 1341–1351. DOI: 10.1111/j.1744-7909.2007.00527.x.

Shao, Yiming; Yu, Xinxing; Xu, Xuwen; Li, Yong; Yuan, Wenxin; Xu, Yan et al. (2020): The YDA-MKK4/MKK5-MPK3/MPK6 cascade functions downstream of the RGF1-RGI ligand-receptor pair in regulating mitotic activity in root apical meristem. In: *Molecular Plant* 13 (11), S. 1608–1623. DOI: 10.1016/j.molp.2020.09.004.

Sheerin, David J.; Menon, Chiara; zur Oven-Krockhaus, Sven; Enderle, Beatrix; Zhu, Ling; Johnen, Philipp et al. (2015): Light-activated phytochrome A and B interact with members of the SPA family to promote photomorphogenesis in Arabidopsis by reorganizing the COP1/SPA complex. In: *The Plant Cell* 27 (1), S. 189–201. DOI: 10.1105/tpc.114.134775.

Shen, Hui; Moon, Jennifer; Huq, Enamul (2005): PIF1 is regulated by light-mediated degradation through the ubiquitin-26S proteasome pathway to optimize photomorphogenesis of seedlings in Arabidopsis. In: *The Plant journal* 44 (6), S. 1023–1035. DOI: 10.1111/j.1365-313X.2005.02606.x.

Shevchenko, A.; Wilm, M.; Vorm, O.; Mann, M. (1996): Mass spectrometric sequencing of proteins silver-stained polyacrylamide gels. In: *Analytical Chemistry* 68 (5), S. 850–858. DOI: 10.1021/ac950914h.

Shin, Dong Ho; Choi, Myoung-Goo; Lee, Hyun Kyoung; Cho, Misuk; Choi, Sang-Bong; Choi, Giltsu; Park, Youn-Il (2013): Calcium dependent sucrose uptake links sugar signaling to anthocyanin biosynthesis in Arabidopsis. In: *Biochemical and biophysical research communications* 430 (2), S. 634–639. DOI: 10.1016/j.bbrc.2012.11.100.

Shin, Dong Ho; Choi, MyungGoo; Kim, Keunhwa; Bang, Geul; Cho, Misuk; Choi, Sang-Bong et al. (2013): HY5 regulates anthocyanin biosynthesis by inducing the transcriptional activation of the MYB75/PAP1 transcription factor in Arabidopsis. In: *FEBS letters* 587 (10), S. 1543–1547. DOI: 10.1016/j.febslet.2013.03.037.

Shin, Jieun; Kim, Keunhwa; Kang, Hyojin; Zulfugarov, Ismayil S.; Bae, Gabyong; Lee, Choon-Hwan et al. (2009): Phytochromes promote seedling light responses by inhibiting four negatively-acting phytochrome-interacting factors. In: *Proceedings of the National Academy of Sciences of the United States of America* 106 (18), S. 7660–7665. DOI: 10.1073/pnas.0812219106.

Shin, Jieun; Park, Eunae; Choi, Giltsu (2007): PIF3 regulates anthocyanin biosynthesis in an HY5-dependent manner with both factors directly binding anthocyanin biosynthetic gene promoters in Arabidopsis. In: *The Plant journal* 49 (6), S. 981–994. DOI: 10.1111/j.1365-313X.2006.03021.x.

Sivitz, Alicia B.; Reinders, Anke; Johnson, Meghan E.; Krentz, Anthony D.; Grof, Christopher P. L.; Perroux, Jai M.; Ward, John M. (2007): Arabidopsis sucrose transporter AtSUC9. High-affinity transport activity, intragenic control of expression, and early flowering mutant phenotype. In: *Plant Physiology* 143 (1), S. 188–198. DOI: 10.1104/pp.106.089003.

Sivitz, Alicia B.; Reinders, Anke; Ward, John M. (2008): Arabidopsis sucrose transporter AtSUC1 is important for pollen germination and sucrose-induced anthocyanin accumulation. In: *Plant Physiology* 147 (1), S. 92–100. DOI: 10.1104/pp.108.118992.

Smith, E. L.; Markland, F. S.; Kasper, C. B.; DeLange, R. J.; Landon, M.; Evans, W. H. (1966): The complete amino acid sequence of two types of subtilisin, BPN' and Carlsberg. In: *The Journal of Biological Chemistry* 241 (24), S. 5974–5976.

Smith, Stephanie; Zhu, Shanshuo; Joos, Lisa; Roberts, Ianto; Nikonorova, Natalia; Vu, Lam Dai et al. (2020): The CEP5 peptide promotes abiotic stress tolerance, as revealed by quantitative proteomics, and attenuates the AUX/IAA equilibrium in Arabidopsis. In: *Molecular & cellular proteomics: MCP* 19 (8), S. 1248–1262. DOI: 10.1074/mcp.RA119.001826.

Soltabayeva, Aigerim; Dauletova, Nurbanu; Serik, Symbat; Sandybek, Margulan; Omondi, John Okoth; Kurmanbayeva, Assylay; Srivastava, Sudhakar (2022): Receptor-like kinases (LRR-RLKs) in response of plants to biotic and abiotic stresses. In: *Plants* 11 (19). DOI: 10.3390/plants11192660.

Somoza, Sofía C.; Sede, Ana R.; Boccardo, Noelia A.; Muschietti, Jorge P. (2021): Keeping up with the RALFs: how these small peptides control pollen-pistil interactions in Arabidopsis. In: *New Phytologist* 229 (1), S. 14–18. DOI: 10.1111/nph.16817.

Song, Wen; Liu, Li; Wang, Jizong; Wu, Zhen; Zhang, Heqiao; Tang, Jiao et al. (2016): Signature motif-guided identification of receptors for peptide hormones essential for root meristem growth. In: *Cell Research* 26 (6), S. 674–685. DOI: 10.1038/cr.2016.62.

Srivastava, Renu; Liu, Jian-Xiang; Howell, Stephen H. (2008): Proteolytic processing of a precursor protein for a growth-promoting peptide by a subtilisin serine protease in Arabidopsis. In: *The Plant journal* 56 (2), S. 219–227. DOI: 10.1111/j.1365-313X.2008.03598.x.

Stadler, R.; Truernit, E.; Gahrtz, M.; Sauer, N. (1999): The AtSUC1 sucrose carrier may represent the osmotic driving force for anther dehiscence and pollen tube growth in Arabidopsis. In: *The Plant Journal* 19 (3), S. 269–278. DOI: 10.1046/j.1365-313x.1999.00527.x.

Stahl, Elia; Fernandez Martin, Angel; Glauser, Gaétan; Guillou, Marie-Charlotte; Aubourg, Sébastien; Renou, Jean-Pierre; Reymond, Philippe (2022): The MIK2/SCOOP signaling system contributes to Arabidopsis resistance against herbivory by modulating jasmonate and indole glucosinolate biosynthesis. In: *Frontiers in Plant Science* 13, S. 852808. DOI: 10.3389/fpls.2022.852808.

Stark, D. M.; Timmerman, K. P.; Barry, G. F.; Preiss, J.; Kishore, G. M. (1992): Regulation of the Amount of Starch in Plant Tissues by ADP Glucose Pyrophosphorylase. In: *Science* 258 (5080), S. 287–292. DOI: 10.1126/science.258.5080.287.

Stegmann, Martin; Monaghan, Jacqueline; Smakowska-Luzan, Elwira; Rovenich, Hanna; Lehner, Anita; Holton, Nicholas et al. (2017): The receptor kinase FER is a RALF-regulated scaffold controlling plant immune signaling. In: *Science* 355 (6322), S. 287–289. DOI: 10.1126/science.aal2541.

Stegmann, Martin; Zecua-Ramirez, Patricia; Ludwig, Christina; Lee, Ho-Seok; Peterson, Brenda; Nimchuk, Zachary L. et al. (2022): RGI-GOLVEN signaling promotes cell surface immune receptor abundance to regulate plant immunity. In: *EMBO reports* 23 (5), e53281. DOI: 10.15252/embr.202153281.

Stephenson, Patrick G.; Fankhauser, Christian; Terry, Matthew J. (2009): PIF3 is a repressor of chloroplast development. In: *Proceedings of the National Academy of Sciences of the United States of America* 106 (18), S. 7654–7659. DOI: 10.1073/pnas.0811684106.

Steyn, W. J.; Wand, S. J. E.; Holcroft, D. M.; Jacobs, G. (2002): Anthocyanins in vegetative tissues: a proposed unified function in photoprotection. In: *New phytologist* 155 (3), S. 349–361. DOI: 10.1046/j.1469-8137.2002.00482.x.

Stintzi, Annick; Schaller, Andreas (2022): Biogenesis of post-translationally modified peptide signals for plant reproductive development. In: *Current Opinion in Plant Biology* 69, S. 102274. DOI: 10.1016/j.pbi.2022.102274.

Stracke, Ralf; Ishihara, Hirofumi; Huep, Gunnar; Barsch, Aiko; Mehrtens, Frank; Niehaus, Karsten; Weisshaar, Bernd (2007): Differential regulation of closely related R2R3-MYB transcription factors controls flavonol accumulation in different parts of the Arabidopsis thaliana seedling. In: *The Plant journal* 50 (4), S. 660–677. DOI: 10.1111/j.1365-313X.2007.03078.x.

Streb, Sebastian; Zeeman, Samuel C. (2012): Starch metabolism in Arabidopsis. In: *The arabidopsis book* 10, e0160. DOI: 10.1199/tab.0160.

References

Stührwohldt, N.; Schaller, A. (2019): Regulation of plant peptide hormones and growth factors by post-translational modification. In: *Plant Biology* 21 Suppl 1 (S1), S. 49–63. DOI: 10.1111/plb.12881.

Stührwohldt, Nils; Bühler, Eric; Sauter, Margret; Schaller, Andreas (2021): Phytosulfokine (PSK) precursor processing by subtilase SBT3.8 and PSK signaling improve drought stress tolerance in Arabidopsis. In: *Journal of Experimental Botany* 72 (9), S. 3427–3440. DOI: 10.1093/jxb/erab017.

Stührwohldt, Nils; Dahlke, Renate I.; Kutschmar, Anke; Peng, Xiongbo; Sun, Meng-Xiang; Sauter, Margret (2015): Phytosulfokine peptide signaling controls pollen tube growth and funicular pollen tube guidance in Arabidopsis thaliana. In: *Physiol Plant* 153 (4), S. 643–653. DOI: 10.1111/ppl.12270.

Stührwohldt, Nils; Dahlke, Renate I.; Steffens, Bianka; Johnson, Amanda; Sauter, Margret (2011): Phytosulfokine-α controls hypocotyl length and cell expansion in Arabidopsis thaliana through phytosulfokine receptor 1. In: *PLOS ONE* 6 (6), e21054. DOI: 10.1371/journal.pone.0021054.

Stührwohldt, Nils; Ehinger, Alexandra; Thellmann, Kerstin; Schaller, Andreas (2020a): Processing and formation of bioactive CLE40 peptide are controlled by posttranslational proline hydroxylation. In: *Plant Physiology* 184 (3), S. 1573–1584. DOI: 10.1104/pp.20.00528.

Stührwohldt, Nils; Scholl, Stefan; Lang, Lisa; Katzenberger, Julia; Schumacher, Karin; Schaller, Andreas (2020b): The biogenesis of CLEL peptides involves several processing events in consecutive compartments of the secretory pathway. In: *eLife* 9. DOI: 10.7554/eLife.55580.

Sun, Jiaqiang; Qi, Linlin; Li, Yanan; Chu, Jinfang; Li, Chuanyou (2012): PIF4-mediated activation of YUCCA8 expression integrates temperature into the auxin pathway in regulating arabidopsis hypocotyl growth. In: *PLoS genetics* 8 (3), e1002594. DOI: 10.1371/journal.pgen.1002594.

Sun, Jiaqiang; Qi, Linlin; Li, Yanan; Zhai, Qingzhe; Li, Chuanyou (2013): PIF4 and PIF5 transcription factors link blue light and auxin to regulate the phototropic response in Arabidopsis. In: *The Plant Cell* 25 (6), S. 2102–2114. DOI: 10.1105/tpc.113.112417.

References

Sun, Tianhu; Zhou, Fei; Huang, Xing-Qi; Chen, Wei-Cai; Kong, Meng-Juan; Zhou, Chang-Fang et al. (2019): ORANGE Represses Chloroplast Biogenesis in Etiolated Arabidopsis Cotyledons via Interaction with TCP14. In: *The Plant Cell* 31 (12), S. 2996–3014. DOI: 10.1105/tpc.18.00290.

Takahashi, Fuminori; Shinozaki, Kazuo (2019): Long-distance signaling in plant stress response. In: *Current Opinion in Plant Biology* 47, S. 106–111. DOI: 10.1016/j.pbi.2018.10.006.

Takahashi, Fuminori; Suzuki, Takehiro; Osakabe, Yuriko; Betsuyaku, Shigeyuki; Kondo, Yuki; Dohmae, Naoshi et al. (2018): A small peptide modulates stomatal control via abscisic acid in long-distance signaling. In: *Nature* 556 (7700), S. 235–238. DOI: 10.1038/s41586-018-0009-2.

Tavormina, Patrizia; Coninck, Barbara de; Nikonorova, Natalia; Smet, Ive de; Cammue, Bruno P. A. (2015): The plant peptidome: An expanding repertoire of structural features and biological functions. In: *The Plant Cell* 27 (8), S. 2095–2118. DOI: 10.1105/tpc.15.00440.

Tian, Miaoying; Benedetti, Brett; Kamoun, Sophien (2005): A Second Kazal-like protease inhibitor from Phytophthora infestans inhibits and interacts with the apoplastic pathogenesis-related protease P69B of tomato. In: *Plant Physiology* 138 (3), S. 1785–1793. DOI: 10.1104/pp.105.061226.

Toledo-Ortiz, Gabriela; Johansson, Henrik; Lee, Keun Pyo; Bou-Torrent, Jordi; Stewart, Kelly; Steel, Gavin et al. (2014): The HY5-PIF regulatory module coordinates light and temperature control of photosynthetic gene transcription. In: *PLoS genetics* 10 (6), e1004416. DOI: 10.1371/journal.pgen.1004416.

Tost, Amalie Scheel; Kristensen, Astrid; Olsen, Lene Irene; Axelsen, Kristian Buhl; Fuglsang, Anja Thoe (2021): The PSY Peptide Family-Expression, Modification and Physiological Implications. In: *Genes* 12 (2). DOI: 10.3390/genes12020218.

Touraine, Brigitte; Vignols, Florence; Przybyla-Toscano, Jonathan; Ischebeck, Till; Dhalleine, Tiphaine; Wu, Hui-Chen et al. (2019): Iron-sulfur protein NFU2 is required for branched-chain amino acid synthesis in Arabidopsis roots. In: *Journal of Experimental Botany* 70 (6), S. 1875–1889. DOI: 10.1093/jxb/erz050.

Truskina, Jekaterina; Brück, Stefanie; Stintzi, Annick; Boeuf, Sophy; Doll, Nicolas M.; Fujita, Satoshi et al. (2022): A peptide-mediated, multilateral molecular dialogue for the coordination of pollen wall formation. In: *Proceedings of the National Academy of Sciences of the United States of America* 119 (22), e2201446119. DOI: 10.1073/pnas.2201446119.

Velasquez, Silvia M.; Ricardi, Martiniano M.; Poulsen, Christian Peter; Oikawa, Ai; Dilokpimol, Adiphol; Halim, Adnan et al. (2015): Complex regulation of prolyl-4-hydroxylases impacts root hair expansion. In: *Molecular Plant* 8 (5), S. 734–746. DOI: 10.1016/j.molp.2014.11.017.

Verma, Vivek; Ravindran, Pratibha; Kumar, Prakash P. (2016): Plant hormone-mediated regulation of stress responses. In: *BMC plant biology* 16, S. 86. DOI: 10.1186/s12870-016-0771-y.

Vie, Ane Kjersti; Najafi, Javad; Winge, Per; Cattan, Ester; Wrzaczek, Michael; Kangasjärvi, Jaakko et al. (2017): The IDA-LIKE peptides IDL6 and IDL7 are negative modulators of stress responses in Arabidopsis thaliana. In: *Journal of Experimental Botany* 68 (13), S. 3557–3571. DOI: 10.1093/jxb/erx168.

Waadt, Rainer; Seller, Charles A.; Hsu, Po-Kai; Takahashi, Yohei; Munemasa, Shintaro; Schroeder, Julian I. (2022): Plant hormone regulation of abiotic stress responses. In: *Nature reviews. Molecular cell biology* 23 (10), S. 680–694. DOI: 10.1038/s41580-022-00479-6.

Wang, Jizong; Li, Hongju; Han, Zhifu; Zhang, Heqiao; Wang, Tong; Lin, Guangzhong et al. (2015): Allosteric receptor activation by the plant peptide hormone phytosulfokine. In: *Nature* 525 (7568), S. 265–268. DOI: 10.1038/nature14858.

Wang, Lei; Einig, Elias; Almeida-Trapp, Marilia; Albert, Markus; Fliegmann, Judith; Mithöfer, Axel et al. (2018): The systemin receptor SYR1 enhances resistance of tomato against herbivorous insects. In: *Nature Plants* 4 (3), S. 152–156. DOI: 10.1038/s41477-018-0106-0.

Wang, Xiaoyang; Zhang, Na; Zhang, Lina; He, Yunxia; Cai, Chao; Zhou, Jinggeng et al. (2021): Perception of the pathogen-induced peptide RGF7 by the receptor-like kinases RGI4 and RGI5 triggers innate immunity in Arabidopsis thaliana. In: *New phytologist* 230 (3), S. 1110–1125. DOI: 10.1111/nph.17197.

Waters, Mark T.; Wang, Peng; Korkaric, Muris; Capper, Richard G.; Saunders, Nigel J.; Langdale, Jane A. (2009): GLK transcription factors coordinate expression of the photosynthetic apparatus in Arabidopsis. In: *The Plant Cell* 21 (4), S. 1109–1128. DOI: 10.1105/tpc.108.065250.

Weise, Sean E.; Liu, Tiffany; Childs, Kevin L.; Preiser, Alyssa L.; Katulski, Hailey M.; Perrin-Porzondek, Christopher; Sharkey, Thomas D. (2019): Transcriptional regulation of the glucose-6-phosphate/phosphate translocator 2 Is related to carbon exchange across the chloroplast envelope. In: *Frontiers in Plant Science* 10, S. 827. DOI: 10.3389/fpls.2019.00827.

Westmuckett, Andrew D.; Hoffhines, Adam J.; Borghei, Atefeh; Moore, Kevin L. (2008): Early postnatal pulmonary failure and primary hypothyroidism in mice with combined TPST-1 and TPST-2 deficiency. In: *General and comparative endocrinology* 156 (1), S. 145–153. DOI: 10.1016/j.ygcen.2007.12.006.

Whitford, Ryan; Fernandez, Ana; Tejos, Ricardo; Pérez, Amparo Cuéllar; Kleine-Vehn, Jürgen; Vanneste, Steffen et al. (2012): GOLVEN secretory peptides regulate auxin carrier turnover during plant gravitropic responses. In: *Developmental Cell* 22 (3), S. 678–685. DOI: 10.1016/j.devcel.2012.02.002.

Wu, Yunzhe; Xun, Qingqing; Guo, Yi; Zhang, Jinghua; Cheng, Kaili; Shi, Tao et al. (2016): Genome-wide expression pattern analyses of the Arabidopsis leucine-rich repeat receptor-like kinases. In: *Molecular Plant* 9 (2), S. 289–300. DOI: 10.1016/j.molp.2015.12.011.

Xi, Lin; Wu, Xu Na; Gilbert, Max; Schulze, Waltraud X. (2019): Classification and interactions of LRR receptors and co-receptors within the Arabidopsis plasma membrane - An overview. In: *Frontiers in Plant Science* 10, S. 472. DOI: 10.3389/fpls.2019.00472.

Xie, De-Yu; Sharma, Shashi B.; Wright, Elane; Wang, Zeng-Yu; Dixon, Richard A. (2006): Metabolic engineering of proanthocyanidins through co-expression of anthocyanidin reductase and the PAP1 MYB transcription factor. In: *The Plant journal* 45 (6), S. 895–907. DOI: 10.1111/j.1365-313X.2006.02655.x.

Xu, Qiyu; Liesche, Johannes (2021): Sugar export from Arabidopsis leaves: actors and regulatory strategies. In: *Journal of Experimental Botany* 72 (15), S. 5275–5284. DOI: 10.1093/jxb/erab241.

Xu, Zhenhua; Mahmood, Kashif; Rothstein, Steven J. (2017): ROS induces anthocyanin production via late biosynthetic genes and anthocyanin deficiency confers the hypersensitivity to ROS-generating stresses in Arabidopsis. In: *Plant & Cell Physiology* 58 (8), S. 1364–1377. DOI: 10.1093/pcp/pcx073.

Yamada, Masashi; Han, Xinwei; Benfey, Philip N. (2020): RGF1 controls root meristem size through ROS signaling. In: *Nature* 577 (7788), S. 85–88. DOI: 10.1038/s41586-019-1819-6.

Yamaguchi, S.; Smith, M. W.; Brown, R. G.; Kamiya, Y.; Sun, T. (1998): Phytochrome regulation and differential expression of gibberellin 3beta-hydroxylase genes in germinating Arabidopsis seeds. In: *The Plant Cell* 10 (12), S. 2115–2126. DOI: 10.1105/tpc.10.12.2115.

Yan, Huiling; Pei, Xiaona; Zhang, Heng; Li, Xiang; Zhang, Xinxin; Zhao, Minghui et al. (2021): MYB-Mediated Regulation of Anthocyanin Biosynthesis. In: *International Journal of Molecular Sciences* 22 (6). DOI: 10.3390/ijms22063103.

Yu, Liangliang; Di, Qi; Zhang, Danping; Liu, Yumin; Li, Xiaolin; Mysore, Kirankumar S. et al. (2022): A legume-specific novel type of phytosulfokine, PSK-δ, promotes nodulation by enhancing nodule organogenesis. In: *Journal of Experimental Botany* 73 (8), S. 2698–2713. DOI: 10.1093/jxb/erac051.

Yu, Liangliang; Liu, Yumin; Zeng, Shuang; Yan, Junhui; Wang, Ertao; Luo, Li (2019): Expression of a novel PSK-encoding gene from soybean improves seed growth and yield in transgenic plants. In: *Planta* 249 (4), S. 1239–1250. DOI: 10.1007/s00425-019-03101-w.

Yuasa, Koji; Toyooka, Kiminori; Fukuda, Hiroo; Matsuoka, Ken (2005): Membrane-anchored prolyl hydroxylase with an export signal from the endoplasmic reticulum. In: *The Plant Journal* 41 (1), S. 81–94. DOI: 10.1111/j.1365-313X.2004.02279.x.

Zeeman, S. C.; Rees, T. Ap (1999): Changes in carbohydrate metabolism and assimilate export in starch-excess mutants of Arabidopsis. In: *Plant, Cell & Environment* 22 (11), S. 1445–1453. DOI: 10.1046/j.1365-3040.1999.00503.x.

Zeeman, S. C.; Umemoto, T.; Lue, W. L.; Au-Yeung, P.; Martin, C.; Smith, A. M.; Chen, J. (1998): A mutant of Arabidopsis lacking a chloroplastic isoamylase accumulates both starch and phytoglycogen. In: *The Plant Cell* 10 (10), S. 1699–1712. DOI: 10.1105/tpc.10.10.1699.

Zhang, Heqiao; Han, Zhifu; Song, Wen; Chai, Jijie (2016): Structural Insight into Recognition of Plant Peptide Hormones by Receptors. In: *Molecular Plant* 9 (11), S. 1454–1463. DOI: 10.1016/j.molp.2016.10.002.

Zhang, Hong; Li, Xiaonan; Wang, Wenping; Li, Huiqiang; Cui, Yanwei; Zhu, Yafen et al. (2022): SERKs regulate embryonic cuticle integrity through the TWS1-GSO1/2 signaling pathway in Arabidopsis. In: *New Phytologist* 233 (1), S. 313–328. DOI: 10.1111/nph.17775.

Zhang, Huan; Hu, Zhangjian; Lei, Cui; Zheng, Chenfei; Wang, Jiao; Shao, Shujun et al. (2018): A Plant Phytosulfokine Peptide Initiates Auxin-Dependent Immunity through Cytosolic Ca2+ Signaling in Tomato. In: *The Plant Cell* 30 (3), S. 652–667. DOI: 10.1105/tpc.17.00537.

Zhang, Huiming; Zhu, Jianhua; Gong, Zhizhong; Zhu, Jian-Kang (2022): Abiotic stress responses in plants. In: *Nature reviews. Genetics* 23 (2), S. 104–119. DOI: 10.1038/s41576-021-00413-0.

Zhang, Luosha; Shi, Xiong; Zhang, Yutao; Wang, Jiajing; Yang, Jingwei; Ishida, Takashi et al. (2019): CLE9 peptide-induced stomatal closure is mediated by abscisic acid, hydrogen peroxide, and nitric oxide in Arabidopsis thaliana. In: *Plant, Cell & Environment* 42 (3), S. 1033–1044. DOI: 10.1111/pce.13475.

Zhang, Ning; Wang, Maike; Huang, Jie; Yang, Leiyun; Wang, Zhixue; Wu, Dianxing; Shu, Xiaoli (2020): MOS1 Negatively Regulates Sugar Responses and Anthocyanin Biosynthesis in Arabidopsis. In: *International Journal of Molecular Sciences* 21 (19). DOI: 10.3390/ijms21197095.

Zhang, Yongqiang; Liu, Zhongjuan; Chen, Yadi; He, Jun-Xian; Bi, Yurong (2015): PHYTOCHROME-INTERACTING FACTOR 5 (PIF5) positively regulates dark-induced senescence and chlorophyll degradation in Arabidopsis. In: *Plant science: an international journal of experimental plant biology* 237, S. 57–68. DOI: 10.1016/j.plantsci.2015.05.010.

Zhang, Zenglin; Liu, Cheng; Li, Kui; Li, Xiaoxu; Xu, Mengmeng; Guo, Yongfeng (2022): CLE14 functions as a "brake signal" to suppress age-dependent and stress-induced leaf senescence by promoting JUB1-mediated ROS scavenging in Arabidopsis. In: *Molecular Plant* 15 (1), S. 179–188. DOI: 10.1016/j.molp.2021.09.006.

Zhao, Chunzhao; Jiang, Wei; Zayed, Omar; Liu, Xin; Tang, Kai; Nie, Wenfeng et al. (2021): The LRXs-RALFs-FER module controls plant growth and salt stress responses by modulating multiple plant hormones. In: *National science review* 8 (1), nwaa149. DOI: 10.1093/nsr/nwaa149.

Zhao, Chunzhao; Zayed, Omar; Yu, Zheping; Jiang, Wei; Zhu, Peipei; Hsu, Chuan-Chih et al. (2018): Leucine-rich repeat extensin proteins regulate plant salt tolerance in Arabidopsis. In: *Proceedings of the National Academy of Sciences of the United States of America* 115 (51), S. 13123–13128. DOI: 10.1073/pnas.1816991115.

Zhao, Dongming; Zheng, Yuxuan; Yang, Lingjun; Yao, Ziyu; Cheng, Jianfeng; Zhang, Fang et al. (2021): The transcription factor AtGLK1 acts upstream of MYBL2 to genetically regulate sucrose-induced anthocyanin biosynthesis in Arabidopsis. In: *BMC plant biology* 21 (1), S. 242. DOI: 10.1186/s12870-021-03033-2.

Zhong, Shangwei; Zhao, Mantong; Shi, Tianying; Shi, Hui; An, Fengying; Zhao, Qiong; Guo, Hongwei (2009): EIN3/EIL1 cooperate with PIF1 to prevent photo-oxidation and to promote greening of Arabidopsis seedlings. In: *Proceedings of the National Academy of Sciences of the United States of America* 106 (50), S. 21431–21436. DOI: 10.1073/pnas.0907670106.

Zhong, Sheng; Liu, Meiling; Wang, Zhijuan; Huang, Qingpei; Hou, Saiying; Xu, Yong-Chao et al. (2019): Cysteine-rich peptides promote interspecific genetic isolation in Arabidopsis. In: *Science* 364 (6443). DOI: 10.1126/science.aau9564.

Zhong, Yingli; Xie, Jiyong; Wen, Suzhen; Wu, Wenwu; Tan, Li; Lei, Mingguang et al. (2020): TPST is involved in fructose regulation of primary root growth in Arabidopsis thaliana. In: *Plant Molecular Biology* 103 (4-5), S. 511–525. DOI: 10.1007/s11103-020-01006-x.

Zhou, Wenkun; Wei, Lirong; Xu, Jian; Zhai, Qingzhe; Jiang, Hongling; Chen, Rong et al. (2010): Arabidopsis Tyrosylprotein sulfotransferase acts in the auxin/PLETHORA pathway in regulating postembryonic maintenance of the root stem cell niche. In: *The Plant Cell* 22 (11), S. 3692–3709. DOI: 10.1105/tpc.110.075721.

Zhu, Tingting; Herrfurth, Cornelia; Xin, Mingming; Savchenko, Tatyana; Feussner, Ivo; Goossens, Alain; Smet, Ive de (2021): Warm temperature triggers JOX and

ST2A-mediated jasmonate catabolism to promote plant growth. In: *Nature communications* 12 (1), S. 4804. DOI: 10.1038/s41467-021-24883-2.

5. Acknowledgement

Zunächst möchte ich mich bei Professor Schaller für die gesamte Zeit an seinem Institut bedanken. Im Jahr 2015 kam ich als junger Bachelorstudent an sein Institut und wurde herzlich aufgenommen. Durch die Bachelorarbeit am Schaller-Institut und die dadurch gewonnene Begeisterung für die molekulare Pflanzenwissenschaft wurde die folgende Richtung meiner Forschung festgelegt. Nach einem erfolgreichen und ebenfalls prägenden Masterstudium in Freiburg, in dem ich die Bedeutung von Phytochromen erforschen konnte, zog es mich für meine Doktorarbeit wieder zurück nach Hohenheim. Erneut wurde ich herzlich im Schaller-Institut aufgenommen und meine Forschung wurde stets durch Rat und Unterstützung von Professor Schaller begleitet. Daher möchte ich mich abschließend nochmals für all die Hilfe über die letzten Jahre bedanken.

Weiterhin möchte ich mich bei Nils Stührwohldt bedanken, der sowohl meine Bachelorarbeit als auch meine Doktorarbeit direkt betreut hat und mir damit maßgeblich bei meinem akademischen Werdegang zur Seite stand.

Als Teil meines Mentorats, Prüfer meiner Dissertation und Institutsleiter meiner Masterarbeit möchte ich auch Professor Hiltbrunner einen besonderen Dank aussprechen.

Ein herzliches Dankeschön geht auch an das gesamte Team des Schaller-Instituts für die Hilfe während der letzten fünf Jahre: Lhana, Annick, Isolde, Rong, Xu, Anne, Ursi, Silke, Marcel, Thomas, Steffi, Bianca, Latisha, Inge, Dagmar, Sally, Fatima, Yalan, Stefanie, Eva, Elsa, Tobi, Jonas, Alissa, Tim, Gerith, Davina, Laura… und viele mehr.

Ich möchte mich auch bei all meinen Kooperationspartnern herzlich bedanken, die meiner Forschung einen anderen Blickwinkel verliehen haben.

Des Weiteren möchte ich mich für die finanzielle Unterstützung des Stipendiums der Landesgraduiertenförderung bedanken.

Zu guter Letzt möchte ich mich noch bei meiner Familie bedanken, ohne die mein gesamter Werdegang nicht möglich gewesen wäre. Vielen Dank für alles.

6. Declaration

Eidesstattliche Versicherung über die eigenständig erbrachte Leistung gemäß § 18 Absatz 3 Satz 5 der Promotionsordnung der Universität Hohenheim für die Fakultäten Agrar-, Natur- sowie Wirtschafts- und Sozialwissenschaften

1. Bei der eingereichten Dissertation zum Thema

“Regulation of stress responses by sulfated peptides in Arabidopsis”

handelt es sich um meine eigenständig erbrachte Leistung.

2. Ich habe nur die angegebenen Quellen und Hilfsmittel benutzt und mich keiner unzulässigen Hilfe Dritter bedient. Insbesondere habe ich wörtlich oder sinngemäß aus anderen Werken übernommene Inhalte als solche kenntlich gemacht.

3. Ich habe nicht die Hilfe einer kommerziellen Promotionsvermittlung oder -beratung in Anspruch genommen.

4. Die Bedeutung der eidesstattlichen Versicherung und der strafrechtlichen Folgen einer unrichtigen oder unvollständigen eidesstattlichen Versicherung sind mir bekannt. Die Richtigkeit der vorstehenden Erklärung bestätige ich. Ich versichere an Eides Statt, dass ich nach bestem Wissen die reine Wahrheit erklärt und nichts verschwiegen habe.

Schuttertal, 04.09.2023

Ort, Datum **Unterschrift**

Lebenslauf

Eric Robin Bühler

Adresse: Am Geisberg 23, Schuttertal, 77978

Geburtsdatum und -ort: 16.08.1993 Lahr

✉ eric.buehler@web.de ☎ 017656919269

Berufliche Erfahrungen

Oktober 2018 – April 2023	Arbeit als **wissenschaftlicher Arbeiter** im Institut für Pflanzenbiochemie von Prof.Dr. Schaller an der Universität Hohenheim.
	Planung und Betreuung von Bachelorarbeiten:
April 2022 – September 2022	• Untersuchungen zur Relevanz von CLEL6 und des RGI5/ SERK3- Rezeptorkomplexes für die Thermomorphogenese in Arabidopsis
Oktober 2021 – Februar 2022	• Analysen zur Bedeutung der LRR Rezeptorkinasen RGI1-5 für die Anthocyanbiosynthese und zur Interaktion zwischen RGI1-5 und dem CLEL6 Peptid aus Arabidopsis
November 2018 –Februar 2019	• Klonierung der extrazellulären Domänen der Rezeptoren At1g29740 und At5g59660 und Tests ihrer Interaktion mit dem CLE-like 9 Peptid
November 2018 –Februar 2019	• Klonierung der Rezeptoren At1g29730 und At3g46370 sowie die Analyse ihrer Interaktionen mit dem CLEL9 Peptid aus Arabidopsis
September 2021 und September 2022	**Leiten eines Kursteils der Pflanzenphysiologischen Übungen** (Biologie Bachelor Modul) an der Universität Hohenheim. Halten der Vorlesung und Durchführen eines PCR Versuches mit Studenten.
08.06.2022	Teilnahme am 1st PhD Symposium in Plant Science an der Universität Hohenheim mit Vortrag
18.-21. Juli 2019	Teilnahme an der Plant Science Student Conference (PSSC) Konferenz am Leibniz Institut für Pflanzenbiochemie in Halle mit Poster:
26.03.2019	Teilnahme an dem 11th *RegioPlantScience* Meeting an der Universität Hohenheim mit kurzer Präsentation und Poster
Sommersemester 2017, M.Sc.	Laborpraktikum am Staatlichen Weinbauinstitut Freiburg
Sommersemester 2014, B.Sc.	Humbold Reloaded Projekt an der Universität Hohenheim: Institut: Systembiologie der Pflanzen (Prof. Dr. Schulze)

August 2013	Ferienjob bei der Albea GmbH
Winter 2009 – Winter 2011	Jugendtrainer beim Hockey Club Lahr
26.04.2010 -30.04.2010	BOGY bei der Tierarztpraxis Dr. Beate Schwing in Friesenheim
17.08.2009 – 04.09.2009 und 02.08.2010 – 20.08.2010	Ferienjob im Lager bei der Herrenknecht AG
19.01.2009 – 03.04.2009	Sozialpraktikum bei der Hausaufgabenhilfe in der Geroldseckerschule Lahr

Sonstige Skills

Sehr gute Englisch Kenntnisse	Verfassen der Doktorarbeit und wissenschaftlichen Veröffentlichungen auf Englisch. Kreieren und Vortragen von diversen englischen Präsentationen auf wissenschaftlichen Tagungen
Sehr gute MS Office und Graphpad Kenntnisse	Analyse von wissenschaftlichen Daten Erstellen von Grafiken für wissenschaftliche Veröffentlichungen und Protokolle

Stipendien und Auszeichnungen

2019 Juli – 2022 Juli	Erhalten des „**Landesgraduiertenförderung**" – Stipendiums der Universität Hohenheim.
18.-21. Juli 2019	Gewinnen des "**Best Poster Awards**" bei der Plant Science Student Conference (PSSC) Konferenz am Leibniz Institut für Pflanzenbiochemie in Halle

Wissenschaftliche Veröffentlichungen

Mai 2023, 1. Autor	"Sulfopeptide CLEL6 inhibits anthocyanin biosynthesis in *Arabidopsis thalia*" im "Plant Physiology" Journal. https://doi.org/10.1093/plphys/kiad316
März 2023, akzeptiertes Manuskript	"A Quick Method to Analyze Peptide-Regulated Anthocyanin Biosynthesis" in Methods in Molecular Biology by Springe Nature
Januar 2021, Co-Autor	"Phytosulfokine (PSK) precursor processing by subtilase SBT3.8 and PSK signaling improve drought stress tolerance in Arabidobsis" im "Journal of Experimental Botany". https://doi.org/10.1093/jxb/erab017

Lebenslauf

Studium

Oktober 2018 bis Oktober 2023	**Promotion an der Universität Hohenheim** Institut Physiologie und Biochemie der Pflanzen (Prof. Dr. Schaller) • Titel der Dissertation: Regulation of stress responses by sulfated peptides in Arabidopsis
Oktober 2015 bis Juli 2018	**Master Biologie an der Albert-Ludwigs-Universität Freiburg:** Abschlussnote: 1,8 Masterarbeit am Institut für Molekulare Pflanzenphysiologie • Titel: Untersuchungen zu den molekularen Mechanismen der Schattentoleranz in *Cardamine hirsuta*
Oktober 2012 bis Oktober 2015	**Bachelor Biologie an der Universität Hohenheim:** Abschlussnote: 2,2 Bachelorarbeit am Institut Physiologie und Biochemie der Pflanzen • Titel: Untersuchungen zur Prozessierung der PSK Vorläuferproteine

Schulausbildung

2000 - 2002	Grund- und Hauptschule Reichenbach
2002 – 2004	Geroldseckerschule Lahr
2004 - 2012	Clara-Schumann-Gymnasium Lahr Abschlussnote: 2,4

Schuttertal, 04.09.2023

Ort, Datum

Unterschrift